WEATHER

WEATHER

STORM DUNLOP

CONTENTS

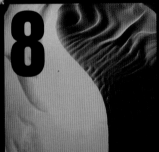

THE EARTH'S CLIMATE IS A KEY PROTAGONIST IN THE STORY OF HUMANITY: our survival as a species has depended on cycles of freeze and thaw, and the ability of our ancestors to adapt to changes in temperature and terrain. The warming and cooling of the planet's oceans and land is vital to its ability to support life. Humans and wildlife migrate from continent to continent, and inhabit different parts of the globe, seeking food and shelter, at the mercy of the weather. As this book shows, life and climate go hand-in-hand, and one of my challenges as a climber is to anticipate changes in the weather and experience the very limits of this partnership. Choosing to press on or retreat in time, before the arrival of a severe storm, can be a life or death decision. The weather not only brings its wind and storms but also affects the nature of the snow cover and the risk of avalanche or stone fall.

The changing weather of our planet – which has experienced Ice Ages and droughts throughout its history – has of course physically shaped the landscape, as well as the life it supports, and the superb images in this book are testament to this. From the wind-sculpted rocks, or yardangs, in the deserts of Asia, to the vast forests of the sub-Arctic taiga, *Weather* presents the evidence of the continuing cycles of winds, rain, cloud and atmospheric optics everywhere in the world. From the immense scale of vast glaciers, a reminder of our once-frozen past, as well as a stark symbol of our precarious future, to the intricacy of beautiful eroded wadis in the Arabian Desert, which seen from above look like tiny rivulets trickling off the landscape, our planet's unique variety of climates is both surprising and spectacular. Of course, dramatic one-off events, such as storms, floods and hurricanes, also bring chaos and devastation in their wake, leaving their own mark on the landscape, as well as on history.

Subtle and sometimes stunning visual manifestations of weather on the other hand, such as auroras and rainbows, provide inspiration for poets, but rarely make front-page news. Long ago thought of as a sign from a powerful deity, the aurora, is, in fact, a dramatic phenomenon brought

FOREWORD

about by charged particles from the Sun entering the upper atmosphere. Similarly, haloes are created by the complex interplay of light reflection and refraction, and while these can be explained, and Storm Dunlop does so succinctly and clearly, still one cannot help but wonder at the power of nature and the apparent intelligence behind such beauty.

Dependent as we are on the elements, so too, we increasingly realise, is the climate affected by humans. In fact, some of those areas which we have prized for their favourable climates, and subsequently populated, are experiencing all too clearly the devastating effects of industry, fuel-dependency and general human over-consumption on our weather patterns. Recent reports suggest that hurricanes in the Gulf of Mexico and surrounding areas are on the increase due to global warming, caused by the burning of fossil fuels, which has the effect of warming our seas. Equally, parts of the Southern Hemisphere are feeling the effects of depleting ozone in the upper levels of the atmosphere. Throughout all these areas, the precious balance of the world's climates looks increasingly precarious.

There is much that we can predict, and the images inside – including sophisticated satellite imagery – are proof of the leaps we have made in even a few years in forecasting the weather. While we do know that our climate is changing, the powerful images here offer a comprehensive picture of not only how far we have come in our understanding of weather but also how far we still have to go.

SIR CHRIS BONINGTON
January 2006

AT FIRST SIGHT, THE INFINITE VARIETY OF CLOUDS APPEARS TO DEFY CLASSIFICATION but, in fact, there are relatively few – ten – major forms (known as 'genera') and these are usually easy to distinguish. Although the general nature of clouds has been known for many centuries, it was not until 1802 that the British pharmacist, Luke Howard, introduced a classification scheme, similar to that used for plants and animals. Although extensively modified subsequently, some of Howard's terms remain in use today.

Meteorologists divide clouds into two broad types: cumulus, or heaped clouds; and stratus, or layer clouds. Cumuliform clouds are usually associated with instability, the condition where a parcel of air, rising in the atmosphere, perhaps as a thermal over warmed ground, remains warmer than its surroundings, and thus continues to rise. By contrast, stratiform clouds are normally associated with stability. If the air in a particular layer is displaced – perhaps being forced upwards by the wind passing over a mountain – when conditions are stable, the air sinks back down towards its original level after passing the obstacle.

Somewhat confusingly to the novice, some clouds exhibit both cumuliform and stratiform characteristics. Stratocumulus and altocumulus, for example, often occur as a layer of cloud that has been broken up into individual cloudlets by shallow convection. An additional group is sometimes mentioned: cirriform clouds. These are high clouds that consist of ice crystals, and thus display a distinct fibrous appearance. Cloud types may also be sub-divided according to their heights. There are three ranges (known as étages): low, medium,

and high. Low clouds are cumulus, stratus, and stratocumulus; medium: altocumulus, altostratus, and nimbostratus. (The last may sometimes descend to the surface.) High clouds are cirrus, cirrostratus, and cirrocumulus. Cumulonimbus may extend through all three levels. Actual heights vary considerably depending on latitude. As a rule, their bases and tops are lower near the poles. It is notable that although meteorologists worldwide use metric units, cloud heights are still specified in feet, because this unit has been adopted internationally by the airline industry.

The tops of giant cumulonimbus clouds in the tropics may reach 60,000 ft (approx. 18 km) or more, whereas toward the poles only very rarely do clouds reach 26,000 ft (approx. 8 km), and are generally much lower. The base of low clouds, by contrast, is relatively constant at about 6,500 ft (approx. 2 km) although it may sometimes descend as low as the surface.

The classification scheme has further sub-divisions: species, which describe general shape and structure; and varieties, which cover cloud transparency and the overall arrangement of individual cloud elements. Most are not shown in detail here, but a few supplementary features have been included because of their striking nature. There are also some clouds that occur far higher in the atmosphere: the rare, but outstandingly beautiful displays of nacreous clouds, which are so striking that they are usually mentioned on the national news; and noctilucent clouds, the highest of all, glowing in the sky at midnight.

1 | BLUE SKIES BECOMING OVERCAST

Nacreous or mother-of-pearl clouds

THESE POLAR STRATOSPHERIC CLOUDS occur at altitudes of 15–30 km/9–19 miles in the lower stratosphere, where normally clouds are rare. Their structure is complex, with initial condensation nuclei of sulphuric acid droplets, on which nitric acid trihydrate is deposited when temperatures drop to -78°C/-108.4°F. The tiny particles (about 1 μm in diameter) are hard to detect, but when they enter a colder region – perhaps as a result of wave motion of the air created by mountains far below – with a rapid drop in temperature below -83°C/-117.4°F, they become coated with a layer of ice. Diffraction of sunlight by the ice-covered particles – still only about 2 μm in diameter – gives rise to pure spectral colours. Slower cooling, as at the onset of winter, produces fewer, but larger particles (about 10 μm), giving rise to clouds that appear white.

Previous pages | A lenticular cloud

A LENTICULAR CLOUD (ALTOCUMULUS LENTICULARIS), photographed with a fish-eye lens over Sandstone Bridge, Alabama Hills, California, USA, at sunset. Lenticular clouds (also known as wave clouds) are produced by the motion of the air above hills or mountains, and may persist for a long time if the wind remains constant in strength and direction. Sandstone Bridge is a natural arch formed by water erosion.

Nacreous clouds

SEEN JUST BEFORE SUNRISE or after sunset, these beautiful clouds hold a hidden hazard. Their tiny particles are a significant factor in the destruction of stratospheric ozone, and the creation of ozone holes. Chlorine from chlorofluorocarbons (CFCs) in the atmosphere becomes active on their surfaces, breaking down ozone. In addition, they remove nitrogen from the air, which would otherwise combine with chlorine to give inert chlorine nitrate. These clouds have become more frequent in recent years but the reasons for this are unknown.

11

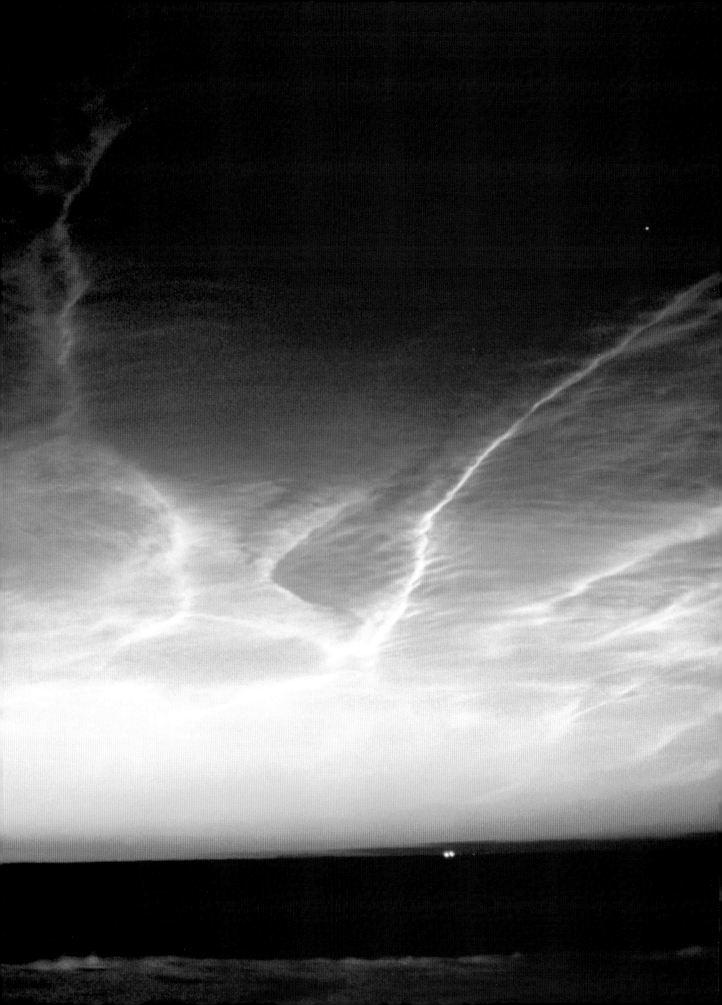

THESE SILVERY-WHITE OR SLIGHTLY YELLOWISH CLOUDS are the highest in the atmosphere, occurring at 80–85 km/50–53 miles, in the mesosphere. They are visible around midnight for a period of about two months around the summer solstice, at high latitude (above about 45° N or S), when the clouds are illuminated by sunlight but the observer is in darkness. They consist of ice particles, formed on freezing nuclei: perhaps consisting of meteoric dust, or clusters of ions produced by cosmic rays. The clouds form a thin sheet, and their billows and other forms are produced by waves in the upper atmosphere. The patches of cloud often move west, while the waves themselves travel in the opposite direction. Noctilucent clouds seem to be becoming more frequent, the reasons for which are not known.

Cumulus clouds

CUMULUS CLOUDS ARE ONE OF A GROUP OF LOW CLOUDS, having bases below about 6,500 ft (approx. 2 km). This photograph, taken in Western Australia, shows how cumulus are separated from another by distances that are roughly as large as the clouds themselves. Depending on the conditions, such an arrangement may persist all day, with individual clouds growing and then dispersing, or the clouds may spread out into a layer of stratocumulus, with narrow gaps between the individual pancake-like cloudlets. With strong solar heating, however, large cumulonimbus clouds are likely to form instead.

Cumulus clouds at sunset

THESE 'FAIR-WEATHER CUMULUS' speak of the end of a fine day. Cumulus clouds typically have rounded tops and flat bases. They are frequently produced by localized convective cells ('thermals') that break away from surfaces that have been heated by the Sun and rise until they reach the condensation level – marked by the flat bases. If the heating is particularly strong they may go on to build much deeper clouds, towering cumulus, or even cumulonimbus clouds. Small cumulus (as here) decay and fade away when solar heating dies down at sunset.

Medium-sized cumulus

MEDIUM-SIZED CUMULUS often show distinct, rounded tops and indications that the convective cells within them are still growing vertically. The cells are leaning slightly towards the right, indicating that the wind is from the left. (Wind speeds generally increase with height, so it is common to see cumulus towers leaning in this way.)

Heavy cumulus clouds

THESE SUBSTANTIAL CLOUDS are the largest form of cumulus (cumulus congestus). The main convective cells still have sharply outlined, rounded tops, which is an indication that freezing has not yet started in the highest regions. Once freezing (known technically as glaciation) begins, the cloud-tops become less distinct and 'softer'. The clouds would then be classed as cumulonimbus. The bases of these clouds are very dark and it is possible that they may shortly give rise to precipitation. Rain from cumulus congestus is common in the tropics and also occurs in temperate regions in summer.

TAKEN AT THE TOP OF MAUNA KEA, HAWAI'I, this photograph shows two types of layer (stratiform) clouds. Close up and in the distance are banks of deep nimbostratus clouds and between them an indistinct, hazy layer of alto-stratus. Seen from below, the nimbostratus would appear as a heavy dark cloud, producing considerable rainfall, and reaching down fairly close to the ground. The base of the altostratus would be higher and, depending on its thickness, the outline of the Sun might just be visible through the cloud, appearing as if through ground glass. A site of a major astronomical observatory, Mauna Kea has prevailing clear, stable conditions, well above the cloud.

Cirrus clouds over a sand dune

THESE CIRRUS CLOUDS OVER KANSAS are less organized than those in the image at right, above, with apparently random swirls of cloud. They are a general indication of fairly quiet conditions, but there are suggestions towards the horizon that they are becoming thicker and covering a greater area of the sky. A low pressure area may be approaching from that direction.

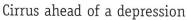

Cirrus ahead of a depression

CIRRUS (ABOVE) ARE THE HIGHEST CLOUDS commonly seen in the atmosphere, and generally have bases at 20,000 ft (approx. 6 km), although they may be only half that height at very high latitudes. They consist of ice particles. Long trails ('mare's tails') as here, indicate strong wind shear at height. When they gradually cover more and more of the sky, all advancing from one direction, they are generally a reliable indication of an approaching depression.

Cirrus floccus

THESE CIRRUS CLOUD TUFTS (BELOW) are seen almost vertically overhead. They show no sign of being drawn out into long streaks and occur as individual tufts (known technically as cirrus floccus). They suggest that the weather is fair and likely to continue so for a while, although the numerous clumps of cloud indicate that there is some instability at height, and this could become significant and assist the growth of cumulonimbus clouds if the latter build up to their height.

21

Cirrostratus clouds and contrails

THE SKY IS COVERED WITH A VEIL of cirrostratus cloud, thickening ahead of the warm front of an approaching depression. When the cirrostratus was thinner, halo phenomena (see page 146) were almost certainly visible. The four persistent contrails (condensation trails) from aircraft are a sign of increasing humidity at height. When the air at altitude is dry, any contrails disperse rapidly. Ahead of a depression, however, they last for a long time and have even been known to completely blanket the sky.

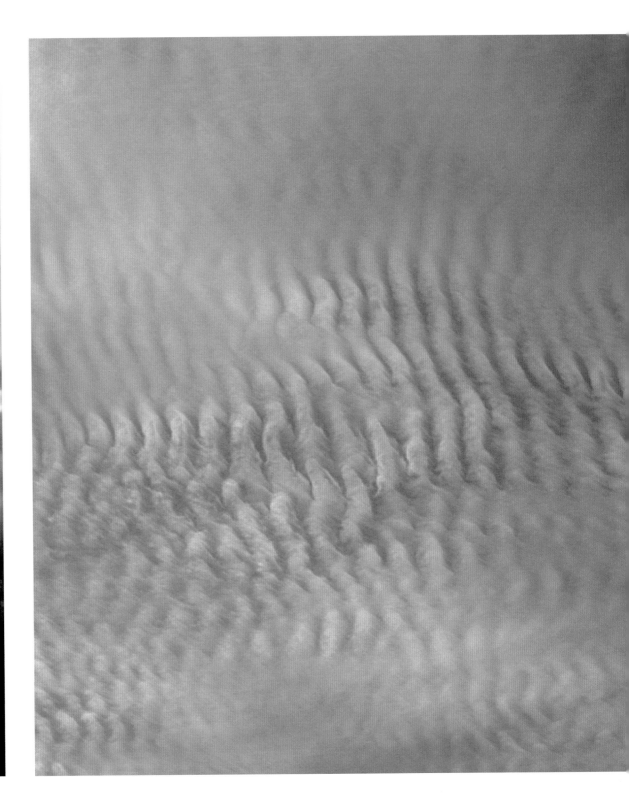

Cirrocumulus billows

LINES OF SMALL, CIRROCUMULUS CLOUDLETS fill the sky. The pattern is commonly known as a 'mackerel sky' – although this term is also frequently applied to altocumulus with a similar appearance. Like the other cirrus clouds, cirrocumulus mainly consist of ice particles. The high and thin cloudlets show no signs of shading, as distinct from lower altocumulus clouds which are thicker with shading on the shadowed sides. The billows are an indication of wind shear at altitude and the billows here lie across the wind. Small patches of cirrocumulus have little significance, but when the whole sky is covered, as here, it usually implies a depression is approaching.

BLUE SKIES BECOMING OVERCAST

Stratus clouds, Ecuador

MULTIPLE LAYERS OF RELATIVELY FEATURELESS STRATUS CLOUD over the Andes in Ecuador. This photograph clearly illustrates how such cloud layers occur at different heights. From the ground, the higher layers would normally be hidden by the lowermost layer. Stratus is a low cloud with a base below 6,500 ft (approx. 2 km) that generally shows few breaks and often shrouds hill- and mountain-tops in cloud; in this way it is identical to fog.

Stratus clouds lapping a line of volcanoes

MERAPI (2,911 M/9,550 FT), SHOWN CENTRE, in central Java, is one of Indonesia's most active volcanoes, but is here showing just a plume of steam. Only its peak is visible, its lower slopes hidden beneath a sheet of stratus cloud. Although carried on the westerly wind, the stratus layer is blocked by the chain of mountainous volcanoes (including Gunung Merbabu, 3,148 m/10,328 ft, top left) that forms the spine of the island of Java, giving some broken cloud and clear skies to the east.

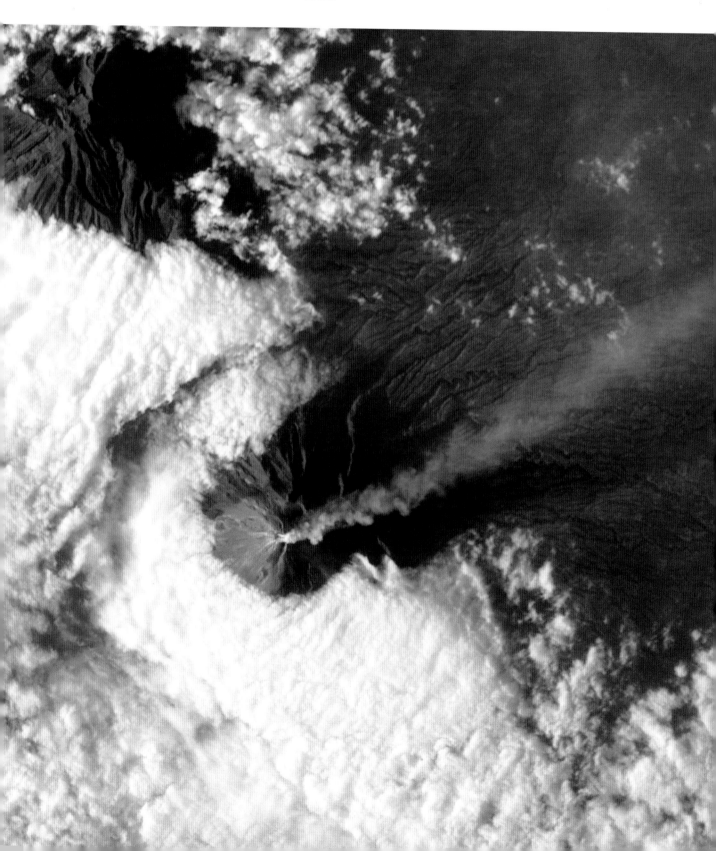

Stratocumulus clouds over the Pacific

THESE CLOUDS MARK THE BOUNDARY of a mass of warm air (top left) that has come from the Californian coast some 580 km/360 miles away. Usually, the clouds form just off the coast. In this instance, however, hot Santa Ana winds from the continental interior have blown the warm air much farther over the ocean before the clouds have formed. To meteorologists' surprise, early satellite images revealed that stratocumulus are the most common clouds, occurring over large areas of the oceans.

Stratocumulus clouds near La Palma

STRATOCUMULUS – LARGE 'PANCAKES' OF CLOUD with narrow clear spaces between them – lie immediately east of La Palma, the largest of the western Canary Islands, in the Atlantic Ocean. North is at top left. The volcanic caldera is clearly visible, and this one is the original feature to have been given that name. ('Caldera' means 'cauldron' in Spanish.) It has been suggested that the western half of the island is unstable and could collapse into the sea, giving rise to a massive tsunami that would swamp Caribbean islands and devastate the eastern coast of the USA, travelling up to 20 km/12 miles inland.

Previous pages | Altocumulus clouds over Mt Erebus

ALTOCUMULUS ARE MIDDLE-LEVEL clouds with bases anywhere between 6,500 and 20,000 ft (approx. 2–6 km), but generally below 12,000 ft (approx. 4 km) at high latitudes. They show distinct shading (unlike the higher cirrocumulus) and frequently accompany lower clouds, such as the cumulus seen here. Mt Erebus on Ross Island in Antarctica is the southernmost active volcano in the world.

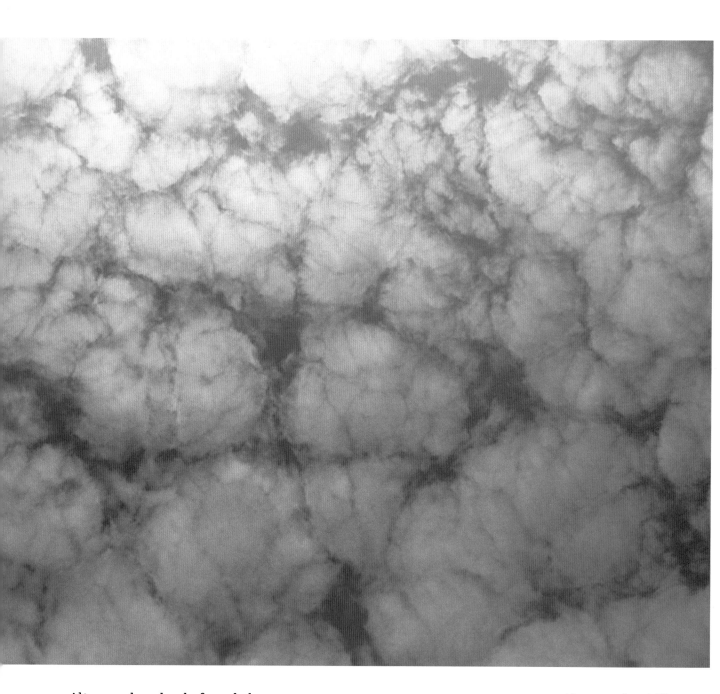

Altocumulus clouds from below

ALTOCUMULUS CLOSELY RESEMBLES THE LOWER STRATOCUMULUS, with 'pancakes' of cloud, separated by narrow gaps of clear air. The two forms are distinguished by their size: altocumulus cloudlets are always less than 5° across, whereas the elements of stratocumulus are larger. Altocumulus may consist of water droplets or a mixture of droplets and ice crystals. Under cold conditions they may give rise to snow, although the particles frequently melt and turn to fine rain before they reach the ground.

Altocumulus billows

BILLOWS SUCH AS THESE (KNOWN AS 'MACKEREL SKY') FORM quite frequently when there is wind shear at height, with one layer of air moving faster, or in a slightly different direction from the layer below it. Altocumulus clouds are defined as having individual cloudlets that are smaller than 5° across – unlike the larger stratocumulus – and larger than 1° across – below which they would be classed as cirrocumulus. Here the clouds merge towards the horizon into a layer of altostratus cloud.

Mature cumulonimbus incus

CUMULONIMBUS CLOUDS FREQUENTLY RISE until they reach the tropopause, an inversion (a region where temperature increases with height). This marks the top of the lowest layer (the troposphere) of the atmosphere. Upward cloud growth is checked, and they spread out into a characteristic anvil-shaped top, and are then known as cumulonimbus incus. The bases may be at the lowest level, but their tops may be as high as 60,000 feet (about 18 km) in the tropics. Here, two individual cells have become glaciated and spread out into anvils. Also visible is the heavy precipitation that is typical of cumulonimbus clouds.

Cumulonimbus cloud with pileus

THIS RAPIDLY GROWING CLOUD has just reached the stage when a cumulus congestus becomes cumulonimbus with the formation of ice crystals in the topmost layers. Two of the active cloud towers have lifted a humid layer of air above the condensation and freezing level, creating a veil of cloud draped over them. This form, known as pileus, is only temporary, because the rising cloud towers will soon penetrate the pileus layer, which will then be incorporated into the active cells by the circulation existing around them.

33

Cumulonimbus clouds at sunset

THIS CLUSTER OF CUMULONIMBUS CELLS over the sea illustrates how such systems develop. Three convective cells ('thermals') have reached the inversion and spread out to form anvils, with the oldest the farthest away. In front of them lies what is known as a 'flanking line' of additional, growing cells, which gradually increase in height towards the body of the storm. If the sea is warmer than the air above it, the cells will continue to grow, even though solar heating has ceased with nightfall.

Cumulonimbus with anvil approaching

CUMULONIMBUS CLOUDS MAY PRODUCE heavy rain, hail, and turn into thunderstorms, regardless of whether they have the typical anvil top. They are the 'showers' frequently described in weather forecasts. The largest clusters of active cells may become multicell storms or even supercell storms, which may persist for many hours, and produce violent weather, including spawning highly destructive tornadoes.

Cumulonimbus clouds seen from space

THIS PHOTOGRAPH, TAKEN FROM LOW EARTH ORBIT, shows cumulonimbus clouds in various stages of growth, with two having developed anvils. From this viewpoint it is easy to see the 'overshooting tops', where the extremely vigorous updraughts have penetrated some distance above the inversion at the top of the clouds. The haze in the background of this photograph of Zaire was caused by the burning of agricultural land.

36

Cumulonimbus cloud and anvil

THIS PHOTOGRAPH OF A SOLITARY CUMULONIMBUS ANVIL was taken from an aircraft flying almost exactly at the level of the trapping inversion. A few wisps of cirrus cloud lie still higher in the atmosphere. Another active cell can be seen rising towards the inversion on the left of the anvil. Although this is the highest cloud in the area, there are signs that other towering clouds will shortly develop.

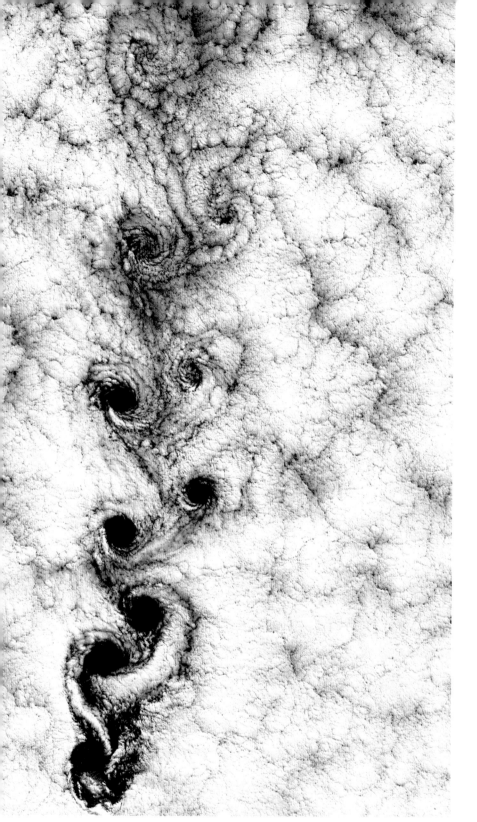

Orographic cloud, Greenland

THIS PHOTOGRAPH SHOWS CLOUD forming where the wind is forcing moist air from the sea to rise over the mountain. In so doing, it reaches the condensation level and produces a layer of cloud, which dissipates to leeward of the mountain where the air descends again. Although the cloud is predominantly stratiform, there are some slight signs of instability in the rounded heads. In the dry Arctic desert, here on Clavering Island in northeast Greenland, rainfall from such clouds is extremely significant and helps to support the hardy vegetation on the mountain slopes.

Orographic cumulus cloud

THESE CLOUDS ARE BUILDING UP in the early afternoon over the southern slopes of Mt Piccolo Altare in the Ligurian Apennines in Italy. The uplift provided by the hills often engenders considerable instability, leading to massive cumulonimbus clouds and accompanying heavy showers. In extreme cases, orographic lifting has led to intense thunderstorms that remain stationary over the peaks and produce destructive flash flooding.

Von Karman vortices in stratocumulus clouds

THESE CLOUD VORTICES IN A SHEET OF STRATOCUMULUS have been created downwind of the peak of Alexander Selkirk Island (1,600 m/5,249 ft), which has disrupted the steady flow. Known as von Karman vortices, this phenomenon occurs at all levels, and is frequently seen associated with mountainous islands such as Svalbard in the Arctic and the Canary Islands in the North Atlantic. Alexander Selkirk Island and Robinson Crusoe Island are part of the Juan Fernandez Islands in the southern Pacific Ocean off the coast of Chile.

Previous pages | Mamma beneath cumulonimbus anvil

THESE DRAMATIC BULGES FREQUENTLY DEVELOP beneath the overhanging anvils
of large cumulonimbus clouds. The top of the anvil radiates heat upwards, cooling the cloud, and cold air sinks beneath it in a form of 'upside-down convection'.
Often mamma take the form of long, irregular tubes, rather than the orderly bulges
seen here. Similar features are found in a number of other cloud types including
cirrus, cirrocumulus, altocumulus, and stratocumulus.

Mamma and crepuscular rays

THE MAMMA IN THIS PHOTOGRAPH are somewhat unusual, in that they have developed beneath a layer of cloud. One portion of the cloud, where the mamma have
formed, appears to be stratocumulus, but other portions are relatively featureless
stratus. Because of the low angle of illumination, with sunlight filtering in from the
edge of the cloud sheet, the mamma are casting distinct shadows (crepuscular
rays) on the underside of the cloud.

Mamma

THE MAMMA HERE SHOW THE TYPICAL appearance of those that develop beneath cumulonimbus
cloud. In this example the mamma are shades of grey as dramatic lighting from low sunlight is
absent. Despite sometimes seeming turbulent and threatening, such formations do not generally herald any dramatic events. They tend to develop at the rear of thunderstorms, and although
they sometimes occur when tornadoes develop, cannot be taken as an indication that such an
event is imminent.

Lenticular cloud

LENTICULAR CLOUDS DERIVE THEIR NAME from their smooth, lens-like shape, and are also known as wave clouds. These particular clouds are altocumulus lenticularis, but similar forms occur in both stratocumulus and cirrocumulus. The designation 'cumulus' might be taken to imply that convection and instability are present, but lenticular clouds are actually a sign of stability (see page 8). They are produced when a humid layer of air is forced to rise above a hill or mountain (in this case Mauna Kea on Hawai'i), reaching the condensation (or freezing) level and giving rise to cloud. Because the air is stable, however, it sinks back to its original level after passing over the peak, and the cloud dissipates. Frequently, however, the air is set into wave motion downwind of the obstruction, and a whole train of wave clouds may be produced.

'Une pile d'assiettes'

FREQUENTLY, LENTICULAR CLOUDS FORM as a stack of clouds, one above the other, and this appearance is known as 'une pile d'assiettes' (French for 'a pile of plates'). They indicate that there is a series of humid layers, separated by drier layers that are all forced to rise and fall in the waves. Depending on the wind strength and direction, and also the nature of the obstacle, more than one stack may occur in the train of waves to leeward of the peaks. These clouds were photographed over the island of South Georgia, in the South Atlantic Ocean.

Jet-stream clouds, Libya

JET STREAMS ARE HIGH-ALTITUDE, HIGH-SPEED RIBBONS of air that develop where there are extreme temperature contrasts in the atmosphere. They may be thousands of kilometres long, hundreds of kilometres wide, and just a few kilometres deep. Cirrus clouds frequently form within them and betray their presence, although at other times the jet streams may be invisible. The clouds often form as billows across the flow, or as long streaks along the stream.

Jet-stream clouds, eastern Canada

FOR A FLOW TO BE CALLED A JET STREAM, the wind speed must exceed 90–100 kph/56–62 mph. The maximum speed ever recorded was 656 kph/408 mph in a jet stream observed over South Uist in the Scottish Outer Hebrides in December 1967. In this view, looking northeast over the Canadian Atlantic seaboard, Cape Breton Island in Nova Scotia is just visible below the cloud at bottom centre.

THE BASIC CONCEPT OF WHAT IS CALLED THE HYDROLOGICAL CYCLE was understood as far back as the 6th Century BC. Thales of Miletus (the famous astronomer also interested in weather patterns) was first to realize that water vapour evaporated from lakes and seas forms clouds, and that it was the same water that fell as rain to replenish rivers, lakes and seas. Slightly later, in the 5th Century BC, another astronomer, Anaxagoras, recognized the process of convection in the atmosphere, and noted that temperatures decreased with height and that the tops of some clouds consisted of ice particles.

These ideas were fundamentally correct, although they were not generally accepted for hundreds of years. We now know that there are two different mechanisms leading to the formation of rain. Cloud droplets are extremely small (0.001–0.05 mm in diameter), so they easily remain suspended in the air, and growth by collision is very slow. When there is strong convection inside deep clouds, however, collisions may eventually produce raindrops, which are generally about 0.5–2.5 mm in diameter. This type of rain is common in the tropics and often occurs in temperate regions in summer time when cumulus clouds may reach high into the sky.

At all latitudes, however, there is another process which occurs when clouds grow high enough for the tem-

2 | SHOWERS WITH SUNNY INTERVALS

perature to drop below 0°C/32°F. The water droplets freeze into ice crystals in the process known technically as 'glaciation'. This is marked by changes in the tops of the clouds, which become fibrous in appearance. The crystals immediately begin to fall under the influence of gravity and collide with other ice crystals or unfrozen water droplets, growing rapidly as they do so. Generally the crystals melt in the lower layers, becoming large raindrops. If the surface temperature is low there may be falls of ice crystals or snow.

Most stratiform clouds, apart from nimbostratus, do not normally produce a great deal of precipitation, but they may give rise to drizzle or, if temperatures are low enough, a fall of tiny ice crystals falling from a higher layer which may 'seed' a lower cloud layer, giving rise to bigger raindrops than might otherwise be expected.

Both condensation and freezing require specific nuclei: tiny solid particles that initiate the processes. Condensation nuclei are present everywhere, but suitable freezing nuclei are often absent, when cloud droplets continue to exist as liquid water at temperatures well below freezing – even down to -40°C/F. Such droplets are found in many clouds, especially at high altitudes, as well as in certain fogs and are said to be supercooled. They freeze immediately when they come in contact with a suitable nucleus or other object (such as an aircraft).

Previous pages | **Heavy shower**

A HEAVY SHOWER FROM A CUMULONIMBUS CLOUD. To the general public, 'showers' are gentle, short-lived falls of rain. To meteorologists, however, showers are falls of rain, snow or hail (often very heavy) associated with cumulonimbus clouds. The grey colour here shows that this heavy shower consists of rain, rather than hail or snow.

Hail and rain over Black Rock Desert, Nevada

VIGOROUS UPDRAUGHTS WITHIN A CUMULONIMBUS CLOUD are required to form hail. Ice crystals that form in the top of a cloud fall into the updraughts, and return to higher layers of the cloud. This may happen several times. They gradually grow as they pass through a region where water droplets are supercooled (i.e. exist well below the normal freezing point). Such droplets freeze immediately on contact, adding to the size of the hail. Eventually the particles become too heavy and fall from the cloud. Here, a white hailshaft is surrounded by grey shafts of rain.

Heavy rain over Scotland

TO METEOROLOGISTS, A SHOWER IS PRECIPITATION from an individual cumulonimbus cloud or cluster of clouds, as distinct from the more-or-less continuous rainfall from frontal cloud. Generally showers cover a restricted area, often no more than 5 km/3 miles across, but multiple-cell storms and supercells cover far greater areas. Rain or hail may be very heavy, especially if (as here over the Sound of Mull) the cloud is moving slowly with the wind. (Note the long straight fallstreaks, which only curve close to the ground, where friction has reduced wind speed.)

ALTHOUGH SHOWERS FROM ISOLATED CUMULONIMBUS CLOUDS affect only a small area at any one time, the amount of precipitation may vary widely. In mature cumulonimbus with strong updraughts, the raindrops tend to be supported by the flow of air, and the amount of rain reaching the ground may be relatively small. As the cloud matures, however, the convection eventually ceases and the updraughts decay. The rush of falling raindrops often produces a powerful downdraught, so a short, intense pulse of rain is accompanied by a powerful gust of wind that spreads out ahead of the cloud.

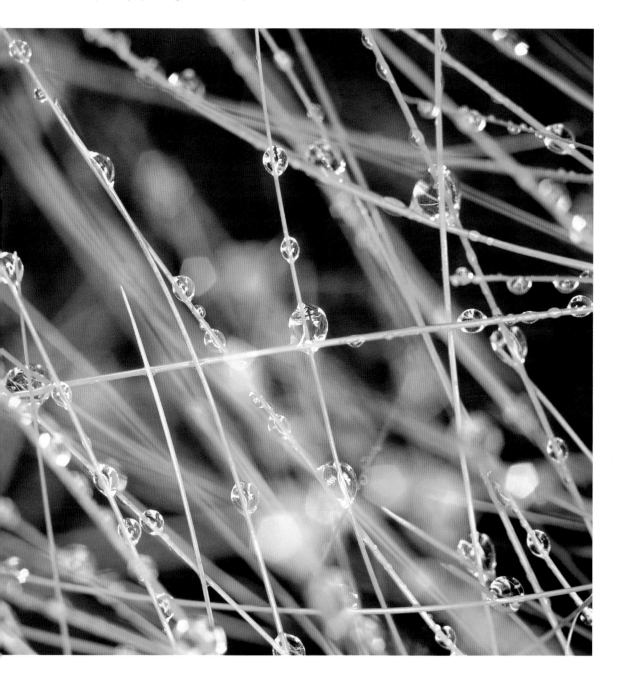

Formation of dew

WHEN THE SKY IS CLEAR AFTER SUNSET and there is little wind, the ground rapidly starts to radiate heat away to space. When the temperature close to the surface reaches the condensation point, water droplets (dew) are deposited on to blades of grass and other surfaces. If the temperature drops even farther, below freezing, water vapour in the air is deposited as ice, giving rise to a coating of hoar frost.

Dewbow

DEW COMMONLY COLLECTS on spiders' webs and, just as with much larger raindrops, the droplets disperse sunlight into a spectrum, similar to a rainbow. This web is vertical, but large dewbows are most frequently seen on extensive areas of grass, covered in a multitude of horizontal webs. Somewhat similar coloured bows may sometimes be seen on tiny water droplets lying on the surface film of ponds and puddles.

Guttation drops on a blade of grass

WATER IS TRANSPORTED WITHIN PLANTS from the roots to the leaves. Frequently, when the temperature drops at night, and particularly when the ground remains warm, excess water is carried to the tips of the leaves where it is unable to evaporate into the air, and forms large guttation drops, which are often mistaken for dew as seen here on a blade of grass.

55

Rime 'feathers'

RIME ICE AT THE SAENTIS METEOROLOGICAL STATION in the Swiss Alps. Supercooled water droplets – i.e. water that is below the normal freezing point – frequently occurs within the atmosphere, especially when there is a lack of suitable freezing nuclei on which they can form as ice. Such supercooled water droplets freeze as soon as they come into contact with a cold surface, giving rise to a deposit of rime. The rime forms on the windward side, so the 'feathers' of rime point into the wind. At high elevations, extremely large deposits, several metres across, often form on exposed surfaces.

Rime ice 'needles'

NEEDLE-SHAPED CRYSTALS of ice such as those seen here on a wire fence, are characteristic of rime formed from supercooled water droplets. All the needles are on one side of the wire mesh. Hoar frost deposits, by contrast, form when the temperature of the air drops below 0°C/32°F and form on all sides of an object.

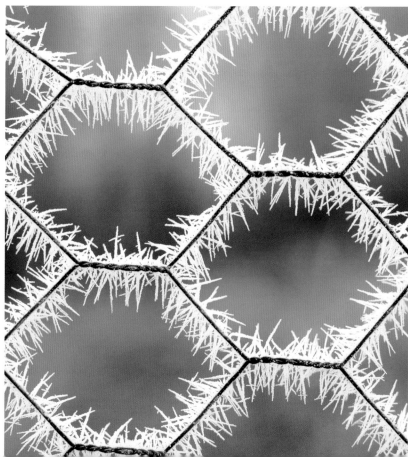

Tree covered in rime

SUPERCOOLED WATER DROPLETS can exist at all levels in the atmosphere, and near the surface often occur as supercooled fog, which freezes instantaneously on contact with trees, shrubs and other surfaces. The surface temperature when this photograph of a Cottonwood tree was taken in the Teton Valley in Idaho was -24°C/-11.2°F.

Frost patterns on window-pane

THESE BEAUTIFUL PATTERNS OF ICE arise from the rapid growth of ice crystals from initial, randomly-scattered, freezing nuclei. The water vapour in the thin layer of air next to the glass is supercooled, and freezes immediately it comes into contact with any existing ice crystal.

Hoar frost

HOAR FROST FORMS IN A SIMILAR WAY to dew, when the temperature falls on still nights with little wind. Instead of condensing, the water vapour freezes on to exposed surfaces to give a characteristically rough deposit. Because there is little air movement, hoar frost does not form the 'feathers' or 'needles' seen with rime, nor the clear layer of ice found with glaze.

Previous pages | A Wyoming snowscape

WINTER SNOW IN UPLANDS AND MOUNTAINS form the principal source of water for the river systems in many parts of the world, including some of the greatest rivers, such as the Missouri/Mississippi system in the USA, and the Ganges and Brahmaputra in Asia.

Extreme rainfall, Indian Ocean

A COMPUTER-GENERATED IMAGE OF THE ISLAND OF REUNION. This isolated oceanic island holds the record for the greatest rainfall in a single day. At Cilaos, 1,870 mm/74 in fell on 16 March 1952. Mountainous islands such as this – the active volcano Piton de la Fournaise is the farther peak – often experience heavy rainfall, leading to dense vegetation, seen here in red.

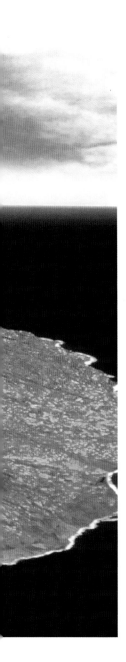

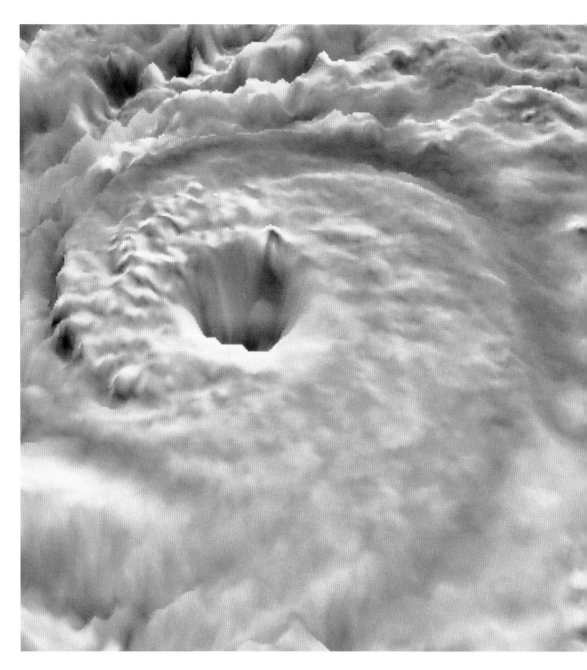

Hurricane Allen

A COMPUTER-GENERATED IMAGE OF HURRICANE ALLEN. Hurricanes are an important source of rainfall for many regions of the world. Despite the damage they may bring, their absence would spell disaster for agriculture and for water supplies in many tropical and sub-tropical countries. Two sets of satellite data were used to create a 3-dimensional model of the storm, which could then be viewed as if from any required angle and altitude, here chosen to display cloud heights.

Eroded wadis, Saudi Arabia

ERODED WADIS NEAR THE LAYLA OASIS IN THE ARABIAN DESERT. The yearly rainfall in this part of Saudi Arabia is only about 100 mm/4 in, but when rain does fall it is unable to sink into the soil but instead gives rise to flash floods that carve deep gullies in the landscape. These wadis may then remain waterless for months or even years, before the next flash flood scours them out again.

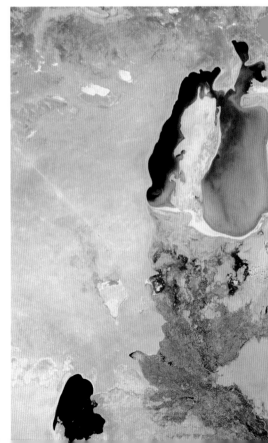

The shrinking Aral Sea

64

THIS SALT-WATER LAKE, LYING IN THE ARID LANDS of Uzbekistan (lower left) and Kazakhstan (upper right) was once the fourth largest lake in the world. Diversion of the waters of the Syr Darya and Amu Darya rivers (upper right and bottom right, respectively) for irrigation has caused the lake (green) to shrink to less than half its former size. Its fishing industry has collapsed, and salt particles, whipped up by fierce winds, are seriously affecting the health of the inhabitants of the region.

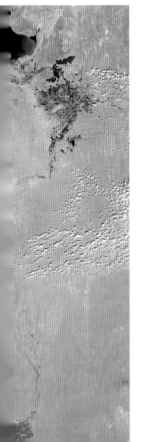
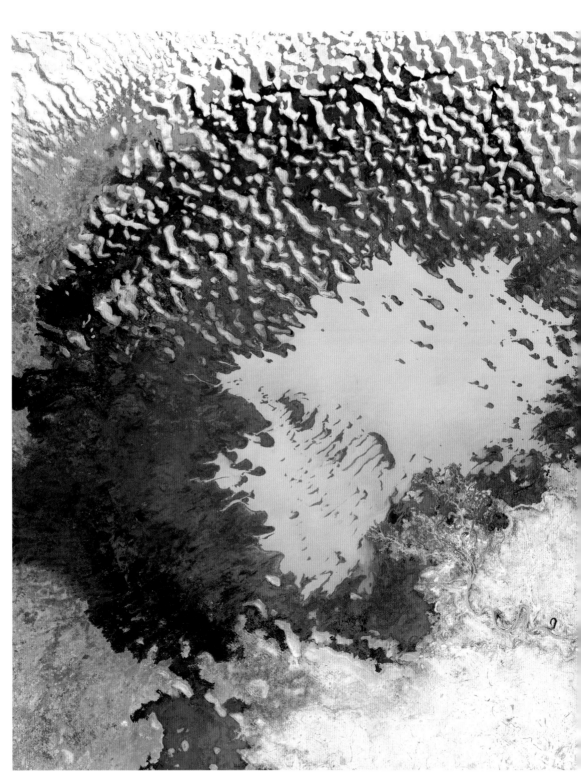

Lake Chad, the fourth largest lake in Africa

THIS LAKE LIES IN THE SAHEL, the border region between the Sahara Desert to the north and the savanna to the south. In 35 years it has shrunk to about one-twentieth of its former size, partly through abstraction of its waters for irrigation, but also because of the recurrent droughts that have been affecting the region, and which appear to be part of long cycles of climate change.

65

Extreme desert: Atacama

THE ATACAMA IS AN EXTREME DESERT with minimal rainfall not only over the lower terrain and the coastal ranges but also in the high Andes. Here two snow-covered volcanoes are seen, but the white areas at upper right are not snow, but salt pans, formed where the meagre precipitation has leached minerals from the volcanic soil, but then evaporated in the high, dry air. The area is almost completely devoid of vegetation, although some scant traces may be seen along some of the streams, where they appear as red tints (caused by use of a satellite image taken in the infrared region).

Forest fires in North America

EVEN RELATIVELY MOIST CLIMATES, such as that of California and Oregon (shown here) may suffer from occasional severe droughts that can provide conditions suitable for highly destructive forest fires. Here, on 29 July 2002, winds from the interior of the continent have desiccated vegetation to such an extent that it was readily set on fire. The smoke plume (yellow/brown in colour) has been swept out over the Pacific. The white lines above the smoke plume are ship tracks, where the exhaust fumes from ships have provided condensation nuclei and created thicker lines of cloud. To the south, the circular pattern of clouds marks the position of tropical storm Elida.

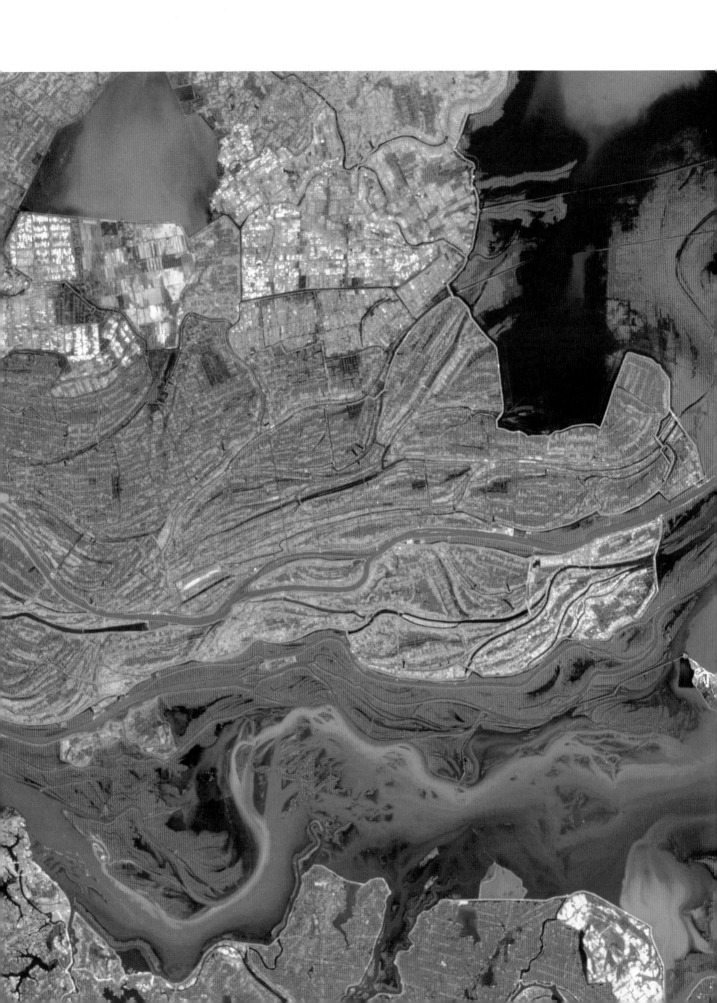

Satellite images of major flooding, China

THESE INFRARED SATELLITE IMAGES OF LAKE DONGTING HU in Hunan Province, China, illustrate the changes during major flooding. In both images, vegetation appears red. In the left-hand image, taken before the floods, the lake appears dark blue. Lighter blue areas are shallow standing water in rice paddies and similar areas. Flood walls are just visible as blue/grey lines.

IN THE RIGHT-HAND IMAGE, taken at the height of the floods, the main lake, now much larger, appears light blue. The flood walls are now clearer to see, and have protected some areas of land. This flood was caused by heavy rainfall in northern China, and was exacerbated by erosion caused by deforestation, which had led to silting-up of areas normally available for water overflow. The floods killed over 3,000 people and destroyed some 5 million homes.

FOG AND MIST – WHICH IS JUST A LESS DENSE FORM OF FOG – are the most obvious causes of poor visibility. They arise whenever the air at ground level cools to the condensation point. Three different types of fog are recognized: upslope fog, when humid air is carried uphill by the wind; radiation fog when the ground radiates heat away to space after nightfall, cooling the layer immediately above it; and advection fog when humid air is carried across a cold surface, such as the sea, or snow- or ice-covered land. This last term is also frequently used for fog that has formed elsewhere and then drifted over a neighbouring area. This frequently happens with fog that forms over the sea and then flows inland.

Various areas of the oceans, where a warm air stream flows over cold water, are particularly prone to fog. The Grand Banks in the North Atlantic are notorious for example, because it is here warm air from the Tropics encounters the cold Labrador Current. Frequent and extensive fogs occur where there are cold oceanic currents and upwelling of cold bottom water from the lowest layers of the ocean. For physical reasons related to the rotation of the Earth, such upwelling takes place along the western side of continents. There are four particular areas where this takes place: two in the Pacific, and two in the Atlantic. Off the coast of South America the Humboldt (or Peru) Current brings frigid water from the Antarctic, and, in the northern Pacific, the California

Current carries cold water south along the coast of California. In the South Atlantic, the Benguela Current lies off the coast of Namibia, and, north of the Equator, the Canary Current (as the name implies) carries cold water past the Canary Islands and the northwestern coast of Africa. All these regions are noted for their large fish stocks, fed by the nutrients brought to the surface by the upwelling.

Visibility may also be affected when tiny dry particles are suspended in the air. This arises naturally where winds raise fine dust from the surface, sometimes in dramatic sand- or duststorms, and also as a result of human activity, often caused by deforestation or (in desert areas) by the destruction of 'desert pavement', the thin, fragile layer that forms over long periods and binds the finer particles into a more resistant layer. Smoke from deliberate or accidental fires also contributes to haze over wide areas of the world.

Air containing a large proportion of smoke and other pollutants is known as smog. Photochemical smog is especially hazardous, and occurs when photochemical reactions, generally powered by sunlight, convert pollutants, particularly vehicle emissions, into highly irritant substances, such as ozone. An inversion can act as a 'lid' trapping the smog layer and preventing its dispersal, as happens with the notorious smogs over Los Angeles, for example. Occasionally photochemical reactions acting on volcanic emissions also create volcanic smog ('vog').

3 LIMITED VISIBILITY

THIS IS AN EXTREME FORM OF VALLEY FOG, created after nightfall when the sky is clear and heat is radiated away to space, cooling the layer of air next to the surface below condensation point. In the case of the Grand Canyon, which lies within the Colorado Plateau, cooling at night produces mountain winds that flow down tributary canyons and add significant contributions of cold air and fog.

Valley fog and light pollution

THIS VIEW OF A LAYER OF VALLEY FOG illuminated by artificial lighting was taken from hills (the Colli Euganei) near Padua in Italy, and illustrates the degree of light pollution that occurs even in a rural area. The colours show the different types of lights: pale yellow, ordinary tungsten-filament lights; orange, sodium vapour; and blue, mercury vapour. The fog is radiation fog that formed over low ground that cooled after sunset.

Radiation fog

RADIATION FOG OCCURS WHEN THE SKY is clear at sunset. The ground rapidly radiates heat that
it gained during the day into space, eventually lowering the temperature close to the surface to
condensation point. Radiation fog collects in low-lying ground and along watercourses. The layer
is usually fairly thin, as in this case photographed in the Great Smoky Mountains National Park,
Tennessee, where the clear sky is visible through the fog.

THE EXTENSIVE ATMOSPHERIC HAZES that occur over China primarily consist of solid particles derived from smoke and dust. Deforestation of vast areas of loess – an extremely fine-grained soil – farther west, has not only caused the whole region to be frequently blanketed by dense haze, but has also served as a source of the massive load of sediment transported by the Huang Ho River which has given rise to its alternative name, the Yellow River. In this image, the bay at centre left is Bo Hai, and North Korea is at upper left.

Sea mist

SEA MIST AND FOG arise when warm air flows across a cold sea. These conditions often produce extensive areas of fog over the sea, and this may be carried by the wind on to neighbouring coasts, giving rise to advection fog. Frequently such fog and mist is confined to the coast because warming over the land causes it to dissipate, giving clear conditions inland. Sea fog or mist tends to persist over the sea, but clears during the daytime over land, returning at nightfall. Often such fogs have local names as here, at Whitby in North Yorkshire, England, where fog from the cold North Sea is known as sea fret or haar.

Fog over the North Sea

THIS FALSE-COLOUR SATELLITE IMAGE of northwestern Europe shows a large bank of fog (pale yellow) over the North Sea. The fog has penetrated inland at various locations along the east coast of Britain. Such conditions arise quite frequently in summer, when there is an easterly flow of warm air from continental Europe that crosses the relatively cold North Sea. High cirrus clouds appear blue and the snow-covered areas of the Alps, Norway and Sweden are clearly visible.

Coastal mist over forested dunes at dawn

ALTHOUGH FOG AND MIST ARE IDENTICAL IN NATURE – both may be described as low cloud – technically, visibility in mist is greater than 1 km/0.6 mile. When visibility is less, it is described as fog. In many areas of the world, advection fog or mist forms a significant (or only) source of water for desert areas, such as the Atacama Desert in Chile, and the Namib Desert in southern Africa. This photograph was taken in the Wilderness National Park, South Africa. In Southern Africa the tenebrionid beetle survives by standing on its head at night, so that dewdrops forming on its body run down to its mouth, providing it with essential moisture.

Arctic sea smoke

ONE TYPE OF FOG OCCURS when extremely cold air moves over relatively warm water. There is rapid evaporation from the water, causing the overlying layer of air to become saturated, producing tendrils of fog that rise from the water. Arctic sea smoke is also known as steam fog or frost fog. It is not confined to polar regions, although this image was obtained near the Ronne Ice Shelf in Antarctica.

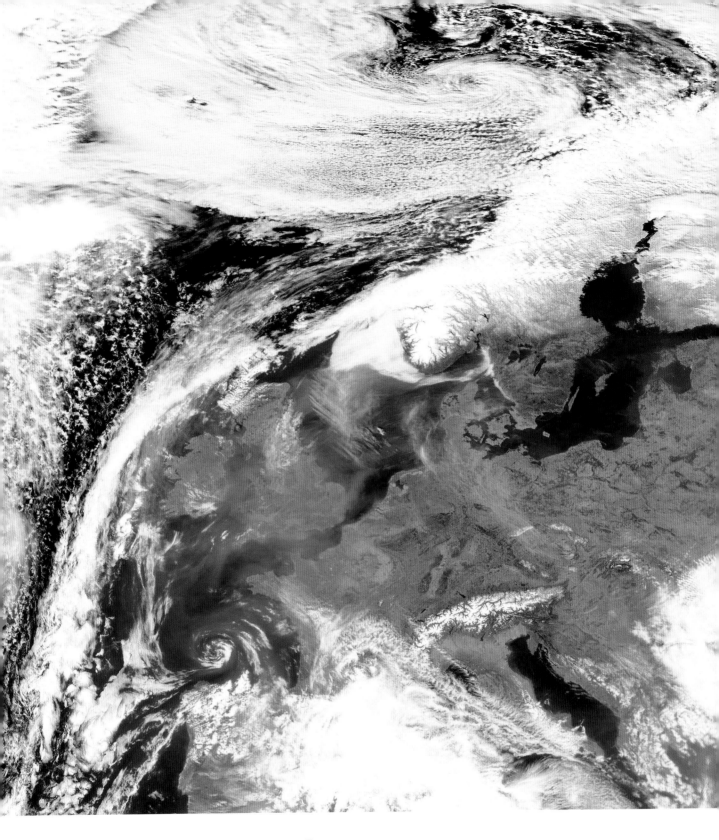

Atmospheric pollution over western Europe

80   GENERAL ANTICYCLONIC (HIGH-PRESSURE) CONDITIONS over Europe, with clear skies and an inversion (see page 33), have trapped a pollution haze in the lowermost layer. It is clearly seen over the Czech Republic, Germany, France, and Great Britain, and it is obvious that it has been spread towards the north and west by the southeasterly winds.

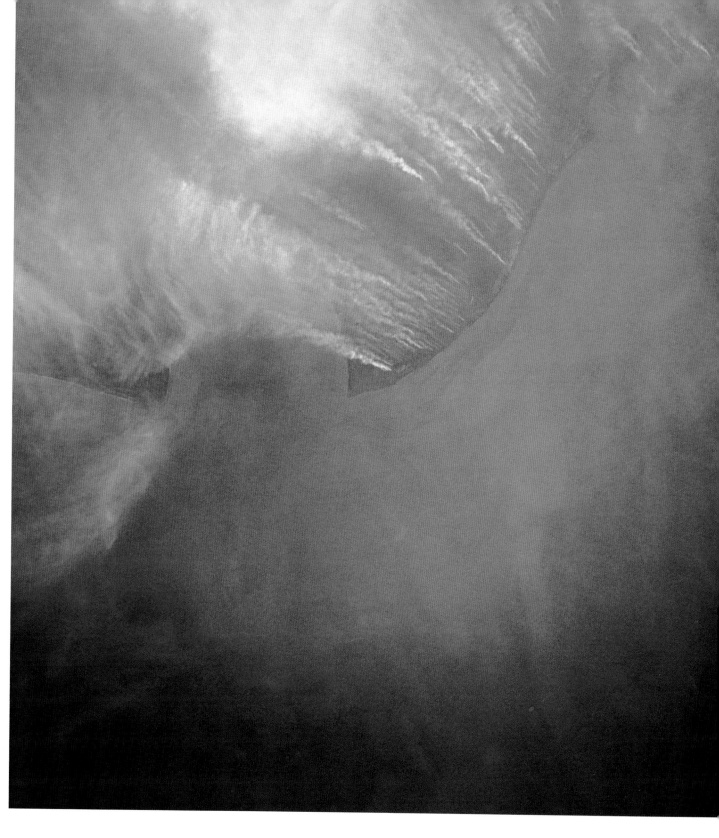

Smoke plumes from land clearance

OBTAINED FROM ONE OF THE SPACE SHUTTLE MISSIONS, this image shows smoke plumes from extensive fires, set to clear marshland in the Tengah region of southern Kalimentan (Indonesian Borneo) to make way for agricultural land, following a surge in population of people emigrating from Java. Such land rapidly becomes infertile, which is having devastating effects on all tropical rainforests. The smoke contributes to extensive hazes that have blanketed most of southeastern Asia in recent years.

81

Saharan sandstorm

A SANDSTORM (WHITE) IS IN PROGRESS over the Djourab region of the Sahara Desert, Chad, at upper centre. The prevailing winds in this region are the Northeast Trades, which blow towards the southwest and have created large linear features on the plains. This view is looking south-west, where the winds are channelled between the dark mountainous regions of the Tibesti Mountains (centre right) and the Ennedi Massif (centre left). The Tibesti Mountains are the high-est point in the Sahara Desert, and their highest peak (Emi Koussi) reaches 3,415 m/11,204 ft.

Sandstorm in Qatar

THIS STRIKING IMAGE OVER QATAR SHOWS a sandstorm sweeping south (to the right) towards Saudi Arabia and the United Arab Emirates. The sandstorm was triggered by strong winds around a deep depression over southwestern Asia. There is a particularly well-marked boundary between the clear, undisturbed desert to the right, and the turbulent clouds of sand and dust to the left.

83

Dust plume

A GIANT PLUME OF SAND AND DUST, originating in Egypt (bottom) has been carried north around the coastline of the eastern Mediterranean, then west and north over Cyprus and the southern coast of Turkey. The densely populated and cultivated areas of the Nile Delta, the El Faiyum depression and along the Nile itself are clearly visible as the dark greenish areas at the bottom of this image.

Duststorm

THE SOIL IN ARID REGIONS such as Arizona, shown here, often has a very fragile surface layer. If this is broken in any way, serious erosion in the form of dust- and sandstorms may occur. Such disturbance, combined with severe drought, led to the disastrous 'dustbowl' conditions over the central plains of the United States in the 1930s.

84

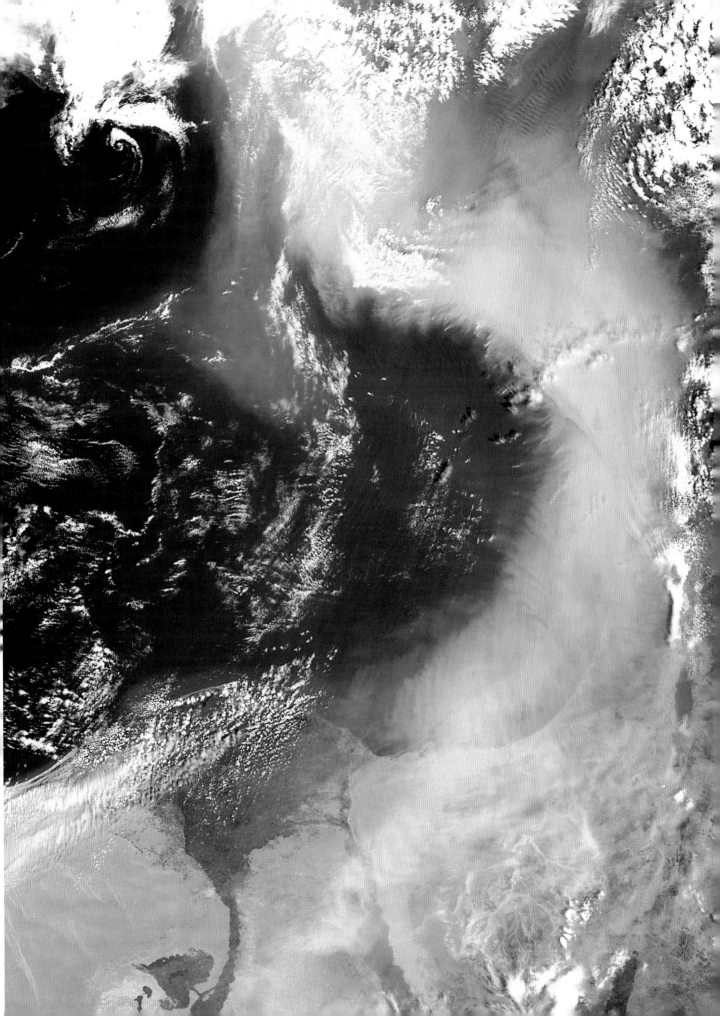

Dust cloud over the Sea of Japan

A VAST PLUME OF DUST from China has passed over Japan and the Sea of Japan and is heading into the Pacific Ocean. Some remnants persist off the coast of North Korea, over the Chosan Man bay on the western side of the Sea of Japan. Also visible is a thin streak of smoke from Mt Oyama, a volcano in central Japan. The dust probably originated in the region of poorly consolidated loess soils in northern China.

Photochemical smog in California

SMOG IS PRODUCED WHEN SUNLIGHT CAUSES photochemical reactions to take place between hydrocarbons and nitrogen oxides, which are primarily derived from vehicle exhausts creating a brown haze of ozone and other irritants. Here, in the case of Los Angeles, California, the problem is exacerbated by the fact that the city is surrounded by hills on all sides except towards the Pacific Ocean. The hills, together with the inversion that is frequently present, trap the polluted air over the city.

IF ASKED ABOUT THE MAJOR DIVISIONS OF THE EARTH'S ENVIRONMENT, most people would mention land in the form of the continents and islands (known technically as the lithosphere), oceans, seas, lakes and rivers (the hydrosphere), and the atmosphere. Only a few would think of the ice sheets of Antarctica and Greenland, the ice caps and glaciers in mountainous regions, and sea ice. Together with permafrost, these form the cryosphere, and play a very large part in determining climate and weather, even though they amount to just 2.8 per cent of the Earth's water and approximately 70 per cent of its fresh water.

The most remarkable feature of Earth's location in the Solar System is that the range of temperatures allows water to exist in all three forms (known as 'phases'): vapour, liquid and solid. The transitions between the phases are particularly important in the atmosphere. The growth of cumuliform clouds, particularly cumulonimbus, may increase dramatically when heat (known as latent heat) is released when water vapour condenses into cloud droplets and water droplets freeze into ice crystals. The formation of ice crystals is an important mechanism in the creation of rain and – although the exact process is still not fully understood – appears to be closely involved with the separation of electrical charges that leads to lightning.

The ice sheets covering Antarctica and Greenland, and the Arctic sea ice, exert a major influence over the

4 | A COLD SNAP

climate and weather of surrounding regions. The vast ice sheets and glaciers that covered large areas of the globe during the last Ice Age are largely responsible for the landscape we see today, particularly in Canada, northern Europe and Asia. During the current period of warming, some glaciers have retreated dramatically by tens of kilometres when compared with their positions a century or more ago. Some, however, are advancing, because the warming trend has produced increased precipitation in their accumulation zones.

Ice shelves and pack ice float, of course, so the increase or decrease in their total area has no direct effect upon sea level. The area covered does, however, have a strong influence on ocean currents and on the weather of surrounding regions. An increase in the amount of fresh water released by melting of parts of the ice caps covering Antarctica and Greenland (especially the latter) could have a major impact on the oceanic circulation, and could lead to a significant deterioration of the climate of western Europe in particular.

On a more immediate level, heavy snowfall and blizzards can cause major disruption to transport and everyday life. Ice storms (more properly described as the widespread incidence of freezing rain which covers any exposed surfaces with large accumulations of glaze) may be highly destructive. The ice storm in Canada and New England on 5–9 January 1998 is a relatively recent example of the ensuing widespread disruption and damage.

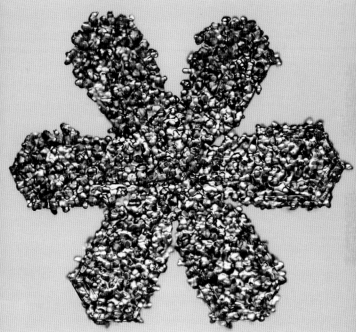

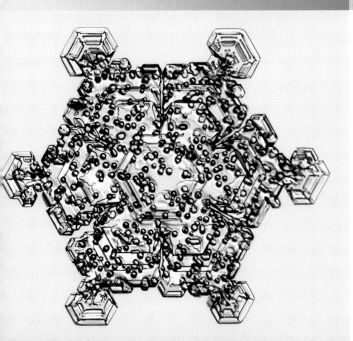

Ice crystals

THESE IMAGES SHOW SOME of the multitude of different, beautiful shapes that may be assumed by ice crystals. An infinite number of other different forms could be included: all exhibiting hexagonal symmetry. Their actual shapes are closely governed by the precise temperature and humidity prevailing when they are formed. Although popularly regarded as 'snowflakes', the latter actually consist of numerous individual ice crystals locked together. Only rarely, under very cold conditions, do single ice crystals like these reach the ground.

Previous pages | Iceberg arch, Greenland

THIS UNUSUAL ICEBERG is locked in frozen sea ice, but at an earlier stage must have been floating freely in the water. An arch was created in the iceberg in the form of a giant hole, the base of which is hidden below the sea surface, which has subsequently frozen. As always, the largest portion of the iceberg is below the surface.

91

Hailstorm, Gillette, Wyoming

HAIL CONSISTS OF SOLID BALLS or pellets of ice. It forms in cumulonimbus clouds with vigorous convection and the consequent strong updraughts. Beginning as raindrops or ice crystals, particles are swept up into the colder regions of the cloud by powerful updraughts, where they collide with supercooled water droplets or other particles. They then begin to fall, but are generally captured once more by the updraughts to complete another cycle. Eventually they become so large that they cannot be supported by the rising air and they fall out of the base of the cloud, sometimes even being 'thrown' downward by strong downdraughts.

Hailstone aggregate

AS IF INDIVIDUAL HAILSTONES were not large enough, they often clump together to form hailstone aggregates, which are even more damaging and dangerous. The largest such aggregates known fell at Hyderabad in India in 1939, with a weight of 3.7 kg/8 lb, and at Yüwu in China in 1902, where the weight was as much as 4 kg/9 lb.

Hailstones

SOME OF THESE INDIVIDUAL hailstones show indications of their internal structure, which consists of concentric layers of opaque and clear ice. When the growing stones pass through cloud layers in which there are relatively warm liquid water droplets, the water spreads out to form a clear layer of ice. In other layers where the droplets are supercooled, they freeze instantly on contact, and air is trapped between the individual tiny droplets, giving rise to an opaque layer. Hailstones may reach 10 cm/4 in across in severe storms. The record weight is 1 kg/2.2 lb, for a single stone about the size of a cantaloupe melon, which fell at Gopalganj in Bangladesh in 1986.

93

Ice crystals under polarized light

THIS MICROGRAPH OF A THIN SECTION of ice reveals a whole range of colours. These patterns appear because different wavelengths of plane-polarized light are rotated by different amounts. The range of colours indicates that there is considerable strain within the ice. This tension arises because the outer layers froze before the inner layers, introducing considerable stress into the crystal structure. It is this internal stress that sometimes leads to hailstones 'exploding' when they hit the ground.

Impending blizzard

THIS CAMP IN GREENLAND is just about to be hit by a heavy snowstorm and high winds. Although the location of the Sun may still be seen, it will soon become completely invisible. Such conditions are popularly called a blizzard but, strictly speaking, the term should be reserved for high winds (Force 7 or more on the Beaufort scale) that are laden with blowing snow raised from the surface, rather than falling snow. In either case, however, visibility is usually extremely low, rendering any travel almost impossible.

Avalanche, west shoulder of Mt Everest

AVALANCHES MOST COMMONLY OCCUR because of the development of a weak layer of snow, known as depth hoar, at some distance below the snow surface. Any slight shock (such as a rock-fall, an earthquake, thunder, or human activity) may be sufficient to set the uppermost layer into motion. This is a powder avalanche, caused by the collapse under its own weight of a deep layer of soft, poorly consolidated snow. Two other, distinct forms of avalanche are the slab avalanche, when a large, relatively coherent layer becomes detached from the underlying layer, and a wet avalanche, where the collapse is initiated by thaw, and which has a high density and may be extremely large.

Glaze or 'black ice'

WHEN SURFACE TEMPERATURES are below freezing, a depression may introduce relatively warm air at a higher level. Rain, falling from the warm layer, freezes on contact with objects on the ground and covers them in a layer of clear ice, known as glaze, and more popularly called 'black ice'. Here, rain falling on a road sign has created a perfect octagon of ice, which has slipped down from its original position.

Rose hips encased in glaze

A LAYER OF GLAZE IS COATING these rose hips after a winter storm that swept across Michigan. It is easy to see how the ice forms a relatively even coating all round the hips and stems of the rose.

97

Hubbard Glacier, Alaska

THIS IS A LARGE GLACIER, and the cruise ship below the centre of the image gives some sense of scale. The heavily crevassed, and weathered upper surface of the glacier is clearly seen. The crevasses, and the effect of the tides, which put considerable strain on the ice, cause large sections to break away to form icebergs.

Ogives on the Gilkey Glacier, Alaska

ANNUAL GROWTH CURVES, OR OGIVES, on a glacier consist of alternating light and dark bands. Raised bands occur where summer warmth has melted some of the surface ice. When the surface re-freezes in winter it is darker because of accumulated dirt and debris. Winter growth produces just clean white ice. The darker ice absorbs more sunlight and partially melts creating shallow troughs in the surface. The ogives are convex downstream, because the sides of the glacier move more slowly than the ice in the centre. The dark streaks parallel to the glacier's direction of movement consist of debris from the rocks at the valley sides.

Crevasses on an ice shelf, Antarctica

THESE DEVELOP WHERE THERE ARE CHANGES in the rate at which the ice is flowing, as a result of friction with the sides of a valley, changes in the width of the valley, or where there are significant changes in the level of bedrock beneath the ice. When the changes are especially pronounced, particularly because of a major, abrupt change in the level of the bedrock, the ice may be so fractured that the area is known as an icefall. On ice shelves, crevasses generally arise because of the continuous flexure of the floating ice due to tides and oceanic swells. This pavement-like pattern is found on the ice shelf in George VI Sound.

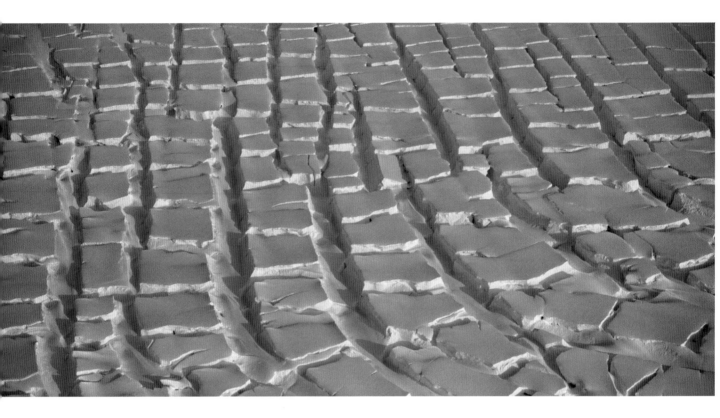

Sastrugi on ice cap, Heer Land, Spitzbergen

SASTRUGI (LEFT) ARE FORMED WHEN hard particles of ice scour the surface, driven by high winds. They carve the surface into a wide range of forms, including grooves that may be as much as 2 m/6.5 ft deep. In this respect they resemble yardangs, (see page 241) the grooves cut into weak sedimentary layers of soil by the action of the wind. Sastrugi usually form in dry snow when temperatures are below -10°C/14°F.

Penitents, Alaska

PENITENTS (also frequently known by the Spanish term 'nieve penitente') arise when the surface of a glacier – which may be snow, firn, or solid ice – becomes heavily fractured. Individual pillars waste away to give rise to tapering spires of ice such as these on the Hole in the Wall Glacier, Alaska .

Meltwater cave, Worthington Glacier, Alaska

102 WATER COMMONLY PENETRATES CREVASSES and accumulates at the base of a glacier, escaping through a cave in the snout. Such streams may be very strong, and carry large quantities of debris. When glaciers retreat, the retreating meltwater caves leave behind snaking lines of rubble, known as eskers. Eskers created by the retreating icecaps of the last Ice Age, and now colonized by vegetation, are often the most prominent features in otherwise flat, eroded landscapes, such as those found in Canada and Finland.

Stranded iceberg, northwestern Greenland

THIS ICEBERG HAS RUN AGROUND in shallow water and the portion normally underwater has been
exposed by low tide, together with the flutes created by its slow melting in sea water. It also shows
that, although most of the mass of an iceberg may be under the surface, it may not be any wider
underwater than it is above the surface.

103

Nunataks, Churchill Mountains, Antarctica

104 NUNATAK IS AN INUIT WORD FOR A MOUNTAIN PEAK that projects above an ice sheet. Most nunataks are angular and jagged because weathering occurs through freeze-thaw cycles that shatter the rock. These nunataks in the Churchill Mountains are part of the Transantarctic Range that borders East Antarctica.

Meltwater pool, Iceland

THIS MELTWATER POOL IS LOCATED ON THE VATNAJÖKULL ICE CAP IN ICELAND. The ice cap covers a large volcanic system, with several active centres. When the volcanoes erupt, parts of the glacier melt and some of the water collects at the surface as meltwater pools, as here, where the black volcanic ash covering the ice may also be seen. Frequently, a large volume of water accumulates beneath the ice, until it suddenly bursts out in a spectacular, and potentially highly dangerous, glacial burst known as a jokülihlaup, creating vast, barren, outwash plains.

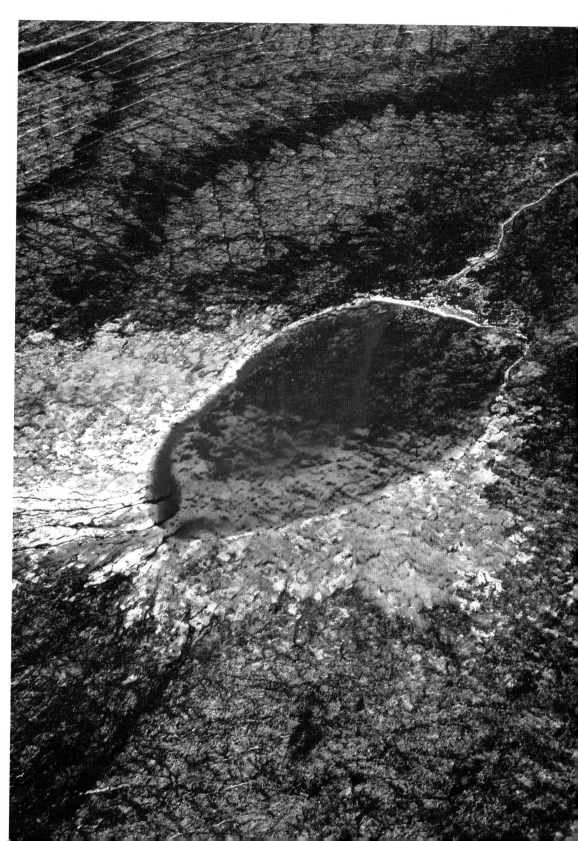

The Tyrolerfjord, Greenland

THIS FJORD IN NORTHEAST GREENLAND consists of a flooded valley cut by an Ice-Age glacier. It illustrates the characteristic U-shaped profile of all glacial valleys, which differs from the distinct V-shape found in valleys cut by rivers. The majority of fjords also have a definite step, or under-water bar, at the seaward end, where the ice causing the erosion lifted away from the bedrock on reaching the sea. The photograph was taken from the summit of Ehrenberg Fjeld in the North East Greenland National Park.

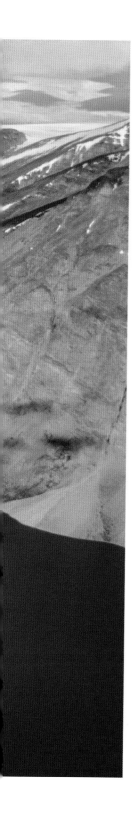

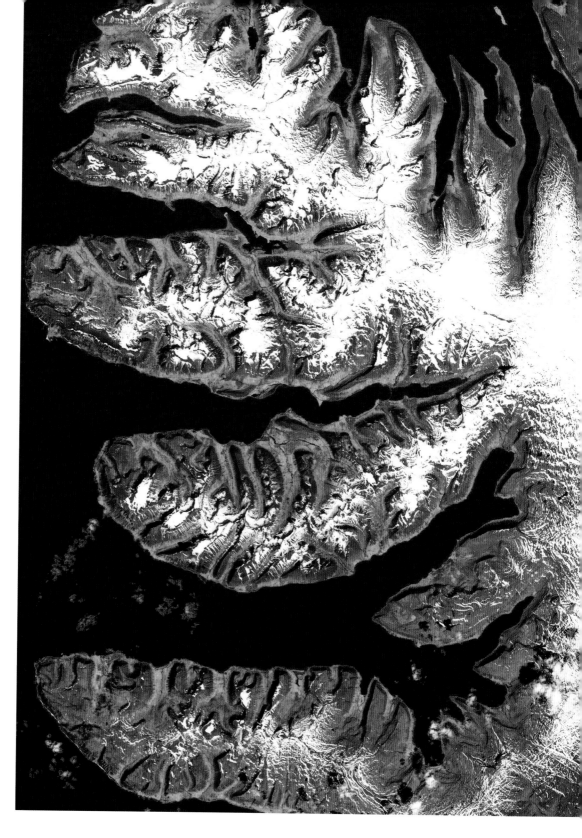

The West Fjords (Vestfirdir) Peninsula, Iceland

THIS IS AN EXTREME EXAMPLE of the type of jagged and indented coastline created by glacial
action and the subsequent flooding following sea-level rise at the end of the Ice Age. A remnant
of the former deep snow and ice cover is still found on the mountain peaks, but the subsidiary
valleys which once held permanent glaciers are now clearly ice-free. The large Arnarfjordur inlet
is at lower left.

107

A COLD SNAP

Islands and sea ice, Trinity Peninsula, Antarctica

THE SNOW-COVERED MOUNTAINS of Trinity Peninsula are top left, Vega Island to the right, and a cluster of smaller islets (some appearing dark with white flecks of snow) are at bottom left. Much of the image is covered by melting shelf ice (bluish tints) which is breaking up into pack ice at the bottom of the image, taken in February, in the southern summer.

Sea ice, Labeuf Fjord, Antarctica

THIS SUMMER IMAGE ILLUSTRATES the complex patterns that occur when sea ice starts to break into ice floes. Initially all the ice was fast ice, that is, ice either frozen to the shore or the sea floor, or else grounded in shallow water. The speckled appearance of the ice is caused by innumerable pools of meltwater on the surface, which appear black. The ice itself is becoming waterlogged, and cracks (white) are beginning to appear, which will eventually widen to create numerous individual floes. Large areas may suddenly break free from the shore or the floor of the fjord and be carried out to sea by the wind.

Ice floes, Arctic Ocean

UNLIKE ANTARCTICA, THE ARCTIC POLAR ICE CAP consists primarily of sea ice, which reaches its maximum thickness and extent in winter. In spring, the ice sheet begins to break up around the edge into innumerable individual ice floes. These drift south, and disperse. The tides appear to play a minor part in the distribution of floes, which are primarily affected by the wind and currents.

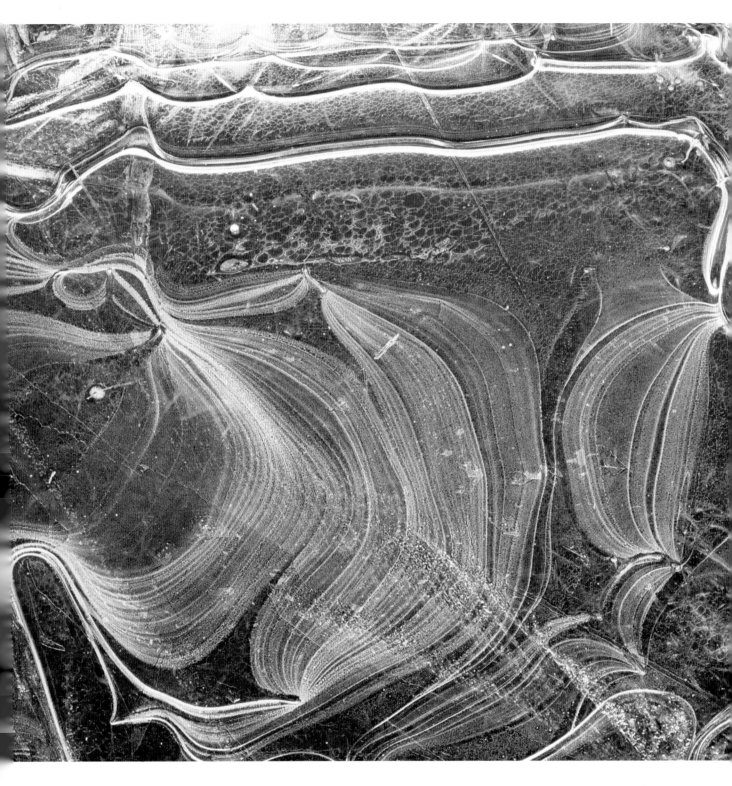

Patterned river ice

THESE SWIRLING PATTERNS betray the changing stresses that occurred as thin ice formed on a river. Similar patterns may often be seen on frozen puddles, where the presence of opaque ice indicates either innumerable tiny bubbles trapped in the ice, or a slight gap between the bottom of the ice and the water, usually because some of the water from the puddle has soaked into the ground after the ice had formed.

EXTREME WEATHER EVENTS CONSTANTLY HIT THE HEADLINES, often prompting the question of whether they are caused by global warming. Although practically all meteorologists and climatologists are certain that global warming is taking place, such is the nature of science that no one can say positively that extreme events have one specific cause. It is often found that a similar natural event occurred in the past. It is also true that the longer you observe any phenomenon, the more likely you are to record extreme events. Weather records obtained with accurate instrumentation date back only about 150 years, so it is not surprising that occasionally events occur that break all previous records.

Modern communications are essentially instantaneous, and this adds to the perception that more extreme events are occurring than happened previously, when news travelled more slowly around the world. But it is precisely this speed of communication and the worldwide distribution of observers and observing stations that are now giving meteorologists the means to make reasonably accurate predictions of extreme weather events. Even though the exact mechanisms responsible for the production of lightning and tornadoes – to give but two examples – are still uncertain, enough is known about atmospheric physics for forecasters to issue 'watches' when severe thunderstorms or tornadoes are possible, and specific 'warnings' when they are imminent.

Similarly, great advances have been made in forecasting the tracks of tropical cyclones (also known as hurri-

canes, typhoons, or cyclones), which are so large, and cause damage and destruction over such large areas that the evacuations of tens or hundreds of thousands of people may be required. Any improvement in the accuracy of such forecasts will not only save lives, but also minimize the disruption that is caused. Tropical cyclones are accompanied by storm surges, but these may also be created by less extreme weather, where the winds accompanying deep depressions may combine with high tides to overwhelm sea defences. Again, it has now become commonplace for such storm surges to be predicted with considerable accuracy.

Major floods that occur through the melting of winter snow pack or excessive winter rainfall are generally predictable some time in advance, although their precise extent and timing depend so greatly upon local factors that the exact details remain uncertain. The prediction of flash floods, which may cause extensive damage and loss of life within a restricted area, is far more difficult to achieve. Disastrous floods such as the Big Thompson Canyon flood in Colorado on 31 July 1976, in which 237 people lost their lives; or the Lynmouth, Devon, flood of 15 August 1952, when over 30 people died, were so localized that no specific warning could be given (nor could one be given nowadays), other than that conditions were such that torrential rain and possible flooding could occur. Now, however, the use of weather radar provides a method of determining where extreme events are occurring in real time, thus giving the emergency services valuable additional information.

5 | WEATHER WARNING

Previous pages | A supercell thunderstorm

WITHIN A SUPERCELL THUNDERSTORM a single massive rotating updraught (known as a mesocyclone) lies in the heart of the cloud. Such supercells persist for many hours and produce violent weather, such as this thunderstorm over Kansas farmland, including extremely heavy rain and hail and frequent lightning. They also often spawn violent tornadoes.

Lightning

TWO OF THE HAZARDS OF SUPERCELL STORMS. Although supercell storms produce numerous lightning strokes and occasional tornadoes, it is rare for both to be captured in a single image, as happened here in Florida. Both are clearly visible, because there is no rain or hail falling to obscure the view. Tornadoes often descend from an area of cloud known as the 'rain-free base'.

Global distribution of lightning

THIS DIAGRAM OF THE ANNUAL FREQUENCY of lightning discharges per square kilometre ranges from red/orange (high) through yellow, green, blue, purple and grey to white (low). For many years, Indonesia was believed to have the highest concentration of lightning strikes, but satellite observations revealed that the region around Kampala in Uganda held the record. Other 'hot spots' occur over Florida, the northwestern Himalaya, and Colombia.

114

Cloud-to-cloud lightning

A LATE-EVENING VIEW of a major thunderstorm over Phoenix, Arizona. Lightning discharges not only occur from cloud to ground, but also within a cloud, between different clouds, and between cloud and clear air. The mechanism by which positive and negative charges become separated is still imperfectly understood, as are the fine details of precisely how discharges take place. Lightning strokes, such as the one shown here, which runs across the base of the cumulonimbus anvil, are known as 'anvil crawlers' to storm chasers.

Cloud-to-ground lightning

LIGHTNING FLASHES NEAR STERLING, COLORADO. The initial channel is created by what is known as a stepped leader, and in most cases this brings negative charge from the cloud down to the ground. The main positive stroke then flows upwards into the cloud. Multiple strokes may follow the same initial channel, consisting of dart leaders from the cloud, and return strokes from the ground. In a small percentage of cases (about 5 per cent) positive charge descends from the cloud, producing a negative return stroke that carries a far greater current.

Lightning strikes near Kourou, French Guiana

TWO METEOROLOGISTS STUDY the distribution of lightning near the Ariane rocket launch site (marked 'ELA') over a period of one month. Knowledge of the likelihood of severe weather and lightning in particular is essential in planning satellite launches. The jagged white line running diagonally across the screen indicates the coastline, and it is noticeable how almost all the lightning strikes occur over land.

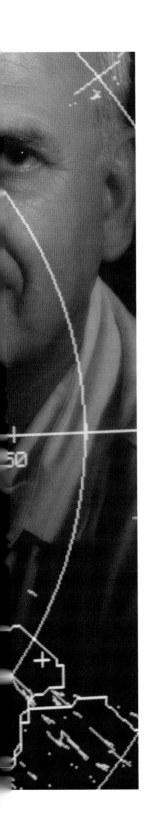

Artificially triggered lightning

AT SOME RESEARCH STATIONS, particularly in France and the USA, lightning is artificially triggered. A rocket is launched into a thunderstorm trailing a fine copper wire. This provides an easy path for the lightning discharge to reach the ground. Such research allows scientists to measure the current, voltage, temperature and other parameters of lightning strokes.

Dust devil

DEVILS (ALSO KNOWN AS WHIRLS) are named after the material that they lift from the surface. Water, snow, and dust devils are fairly common. They generally arise either because the wind is funnelled through a gap and sets up a whirl, or through differences in the roughness of the ground. When there is heating of the surface and consequent convection, dust devils may become stronger than the other types, and lift considerable quantities of dust and fine sand into the atmosphere giving rise to a visible funnel. Devils (whirls) are not directly related to the more destructive tornadoes.

Giant waterspout

WATERSPOUTS AND THE RELATED LANDSPOUTS – a term that has only recently been introduced – have a different formation mechanism from whirls and tornadoes. They arise when there is strong convection within a cumulonimbus cloud and powerful up- and down-draughts. A funnel cloud extends downwards until it reaches the sea. Note the characteristic 'bush' of spray at the base of the spout seen here in the North Atlantic, south of Bermuda.

Steam or vapour spout

AT LOCATIONS (SUCH AS HAWAII) where molten lava reaches the sea, it creates large amounts of steam, which may form localized cloud in the form of a steam or vapour spout. Occasionally, conditions may be right, with strong updraughts and humid air, for the formation of a rotating funnel, within which the reduced pressure causes the vapour to condense. Such steam spouts are normally very short-lived, lasting no more than a few minutes.

Gustnado, New Mexico

A GUSTNADO is a relatively weak whirl that develops from the outflow that occurs along the gust front from a thunderstorm, a line squall, or other storm system. The mechanism by which it forms is similar to that of a dust devil or a landspout, rather than a true tornado, but it may still be strong enough to cause some damage.

123

Supercell thunderstorm

WITHIN AN EXTREME THUNDERSTORM known as a supercell, such as this one in Nebraska, there is a mesocyclone, an exceptionally large rotating updraught that extends to the top of the cloud. This giant updraught is well separated from the accompanying downdraughts and this leads to the long life of supercell storms (six hours or more). Supercells produce high winds, torrential rain, heavy hail, and persistent lightning. Sometimes the rotation extends downwards towards the ground and gives rise to one or more highly destructive tornadoes.

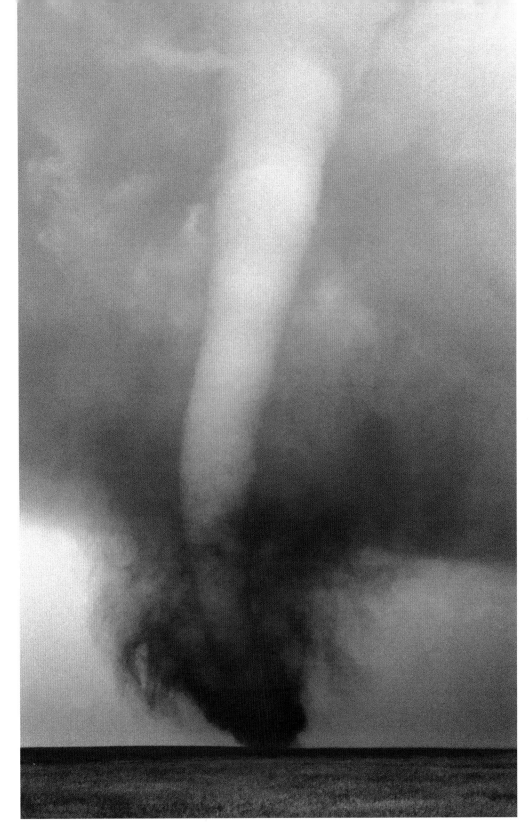

Tornado outbreak, South Dakota

THIS TORNADO, NEAR MANCHESTER, South Dakota, 24 June 2003 was just one in the greatest out-break of tornadoes recorded in the state in one day. The violent, rotating column of air in a torna-do may have a diameter of 100–2,000 m/328–6,562 ft. The highest recorded wind speed of 512 kph/318 mph was measured near Oklahoma City on 3 May 1999. A typical duration is about 15 minutes, but major supercells may spawn a succession of tornadoes over a period of several hours.

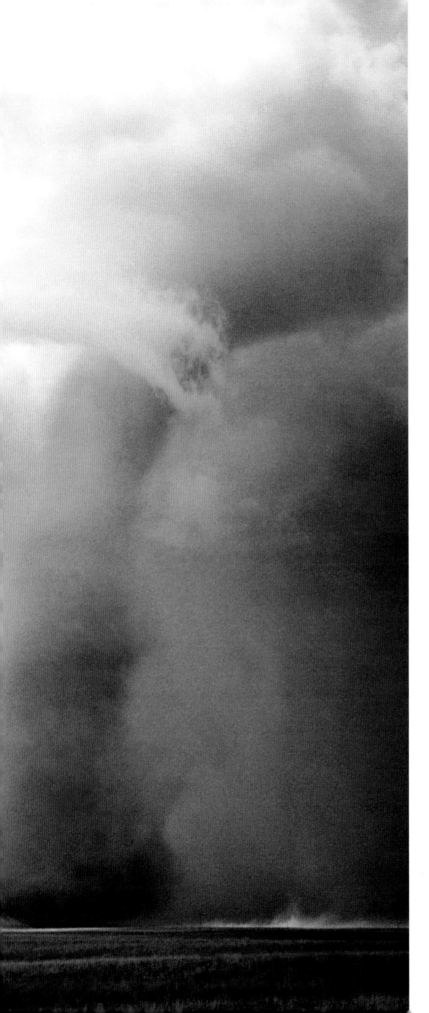

Tornado formation
and development

THESE IMAGES ARE TWO VIEWS OF A TORNADO moving
across Kansas farmland. Major tornadoes form within an
area beneath supercell storms known as the rain-free
base, which is the site of the most powerful updraught, or
from a wall cloud, which is an even lower region that often
exhibits strong rotation. Indications of the wall cloud may
be seen in both photographs.

THE FIRST PHOTOGRAPH (far left) shows the low pressure
in the centre of the tornado causing water vapour to condense, resulting in a funnel cloud hanging from the cloud-
base which is clearly visible. This condensation funnel is
often obscured by the material raised from the surface. In
weak tornadoes, such as those that form beneath non-
supercell storms, the presence of such a debris cloud
may be the only evidence for a tornado, with no funnel
cloud present.

ONCE CONTACT HAS BEEN MADE with the ground, the
diameter of the tornado may expand greatly – as seen in
the photograph near left – but the footprint may vary considerably during a tornado's lifetime, both because of the
changing altitude of the ground, and because the
strength of the vortex fluctuates.

127

Hurricane Mitch

THIS COLOURED THREE-DIMENSIONAL computer image of Central America and the highly destructive Hurricane Mitch is based on satellite data for 26 October 1998, three days before the hurricane made landfall on Honduras. Honduras itself is almost completely hidden by the shield of cirrus cloud above the hurricane. It produced winds of over 320 kph/199 mph and torrential rain that caused flooding and landslides. Over 11,000 people were killed and damage was estimated at US $5 billion.

Multiple-vortex tornado, South Dakota

THIS MULTIPLE-VORTEX TORNADO was one in the major outbreak in South Dakota on 24 June 2003. In major tornadoes, particularly those with an extremely large footprint, subsidiary whirls, known as suction vortices, may form around the base to create a multiple vortex tornado. Tornadoes tend to leave evidence of their passage in the form of swirls on the ground, and the secondary vortices may sometimes be revealed by the small, tighter marks that they produce.

Tropical depression from space

TROPICAL CYCLONES (also known as hurricanes, typhoons, or cyclones, depending on where in the world they occur) develop in a series of stages. Beginning initially as tropical disturbances – areas of low pressure and organized convection – they may develop into tropical depressions, like the one shown here off California. These consist of a series of cloud bands surrounding a low-pressure centre. When the circulation becomes even stronger and more organized, the system may be classified as a tropical storm, the final stage before the development of a true tropical cyclone.

Tropical storm Iniki

THIS STORM ORIGINATED AS HURRICANE INIKI, which caused serious damage on the Hawaiian island of Kauai in September 1992. Winds in the hurricane then exceeded 240 kph/149 mph, and generated waves of over 5 m/16.4 ft in height. When the hurricane passed over Hawaii and into the cooler waters of the North Pacific, it began to weaken and was downgraded to a tropical storm seen here from space. This weakening is shown by the lack of strong convection and thunderstorm activity near the centre of the cloud spiral, and by the breaks in the cloud cover farther out.

Tropical cyclones and tropical storm

THESE TROPICAL CYCLONES and a tropical storm can be seen over the Indian Ocean. The coast of Madagascar appears at far left. The systems visible here, in February 2003, are (from left to right): tropical cyclones Gerry and Hape, tropical storm 18S, and tropical cyclone Fiona. Tropical cyclones form only over tropical oceans where the sea surface temperature exceeds 27°C/80.6°F. Most systems then track westwards until they either cross onto land or move over cooler waters. In both cases their energy source is cut off, and they gradually decay.

132

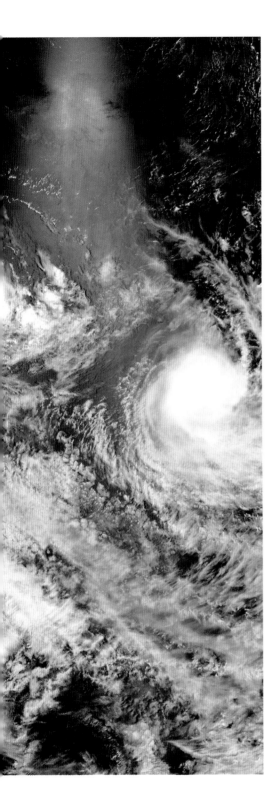

Typhoon Violet over Japan

SOUTH KOREA APPEARS AT TOP LEFT in this picture: a representation of wind speeds and directions in typhoon Violet as calculated from satellite data. Wind direction is indicated by the arrows and wind speed by the colour, from blue (2 m/s/6.6 ft/s) to red (more than 20 m/s/65.6 ft/s). The winds spiral round a central clear, calm eye where the air pressure is extremely low. Typhoon Violet moved over Japan on September 21/22 1996, causing a number of deaths and considerable damage.

Eye of Hurricane Floyd

INSIDE THE EYE OF HURRICANE FLOYD, SEPTEMBER 1999. This view from the cockpit of a Hercules plane, flown by the US Air Force Hurricane Hunters team, shows the towering eyewall. The plane is equipped to measure meteorological parameters, such as wind speed, humidity and atmospheric pressure. Each crew takes a 12-hour flight gathering data from the storm, twice crossing the eye, which involves four passes through the violent cumulonimbus clouds in the eyewall. The data are relayed to base, where computer models help predict the storm's future direction and allow authorities to implement damage-limitation procedures such as evacuation. Hurricane Floyd caused 57 deaths and damage costing US $1.3 billion.

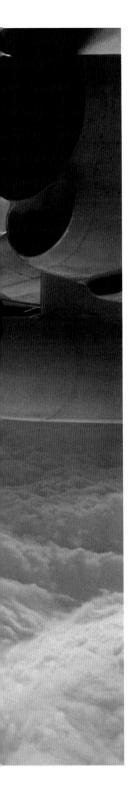

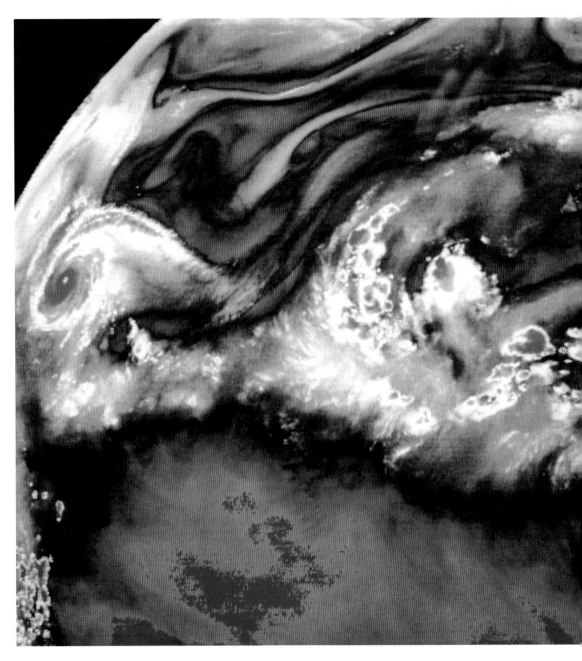

Water-vapour image: Hurricane Floyd

THIS IMAGE FROM THE METEOSAT GEOSTATIONARY SATELLITE was obtained in a spectral band chosen to detect the concentration of water vapour in the atmosphere. Hurricane Floyd (the blue spiral, centre left) lies over the Atlantic. This image was used by the US Air Force Hurricane Hunters team to plan their flight path through the hurricane.

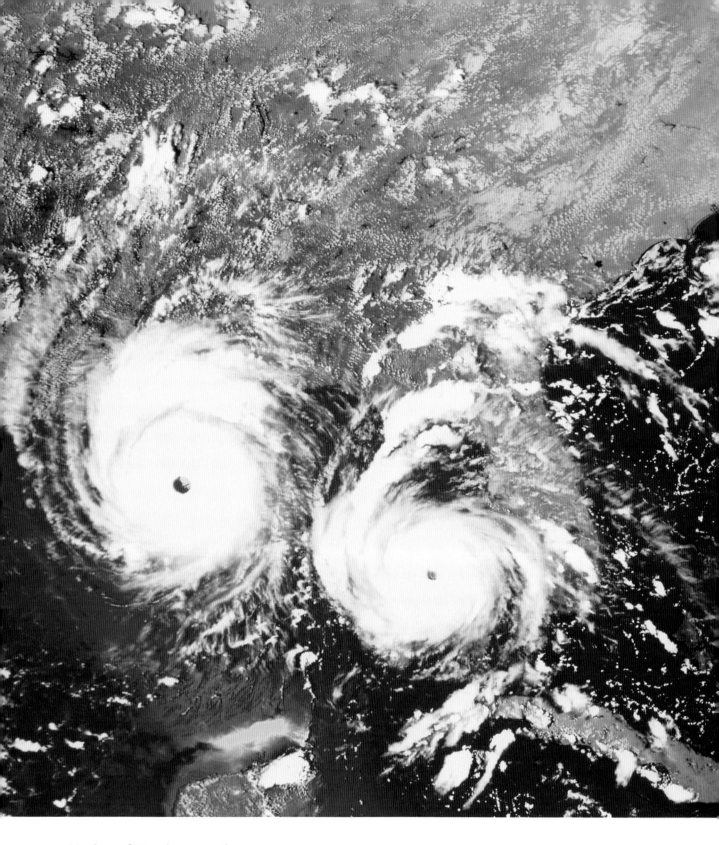

Motion of Hurricane Andrew

A COMPOSITE IMAGE SHOWS the motion of Hurricane Andrew over three days (23–25 August 1992). Andrew, a category 4 event, had winds of 230 kph/143 mph and a central pressure of 922 millibars. Only Camille in 1969 and an unnamed hurricane in 1935 had lower central pressures (909 and 892 millibars, respectively). Andrew was also the costliest hurricane up until then, causing damage amounting to over US $26 billion, more than three times that caused by Hugo in 1989.

Cloud shield and eye of Typhoon Fefa

THIS VIEW FROM THE SPACE SHUTTLE was taken in August 1991 when the tropical cyclone was 1,000 km/621 miles east of Taiwan. The eye, where air descends in the centre of all tropical cyclones, is particularly well-defined, with so little cloud that the sea surface is actually visible. The wall of the eye extends from about 1 km/0.6 mile above the sea to well over 14 km/8.7 miles.

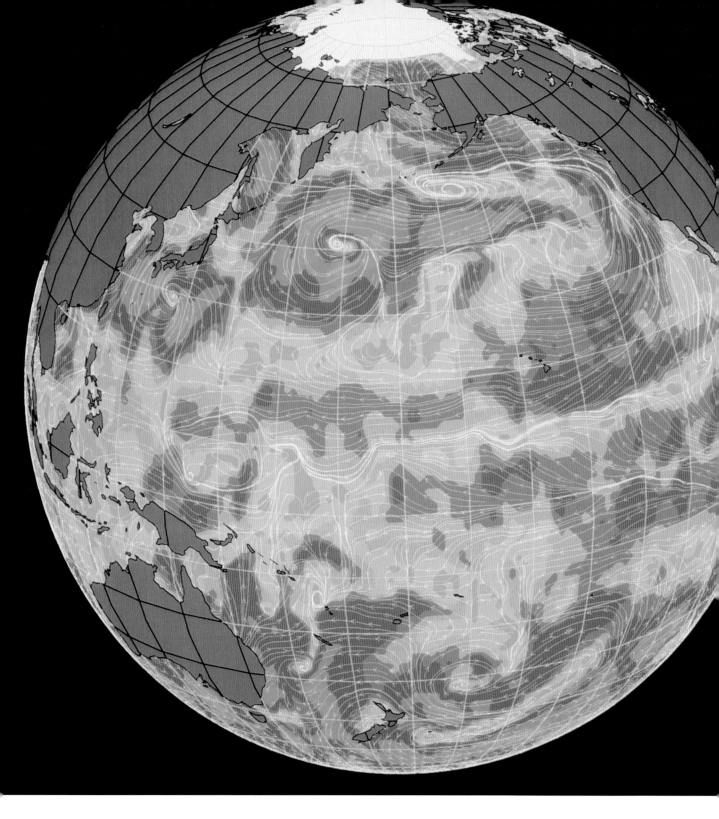

Typhoons near Japan

SYNOPTIC VIEW OF WINDS over the northern Pacific Ocean showing two typhoons near Japan. Wind speed is shown by colours (blue lowest, yellow highest), and wind direction is shown by the white streamlines. Land areas are grey. Typhoon Violet is the system just below Japan (far left) and Typhoon Tom is the larger system at upper centre. Because of the size of the Pacific Ocean and the long paths possible over warm waters, typhoons may become gigantic and more powerful than systems over other oceans. Such 'supertyphoons' include Typhoon Tip in 1979. It reached a diameter of 1,680 km/1,044 miles and had the lowest central pressure ever recorded: 870 millibars on 12 October, when west of Guam.

Radar tracking of hurricanes

THIS PHOTOGRAPH SHOWS a radar screen on board a Hercules aircraft 'Hurricane Hunter' about to penetrate the eyewall of Hurricane Floyd in September 1999. The clear eye of the hurricane is readily visible on the screen. The radar pulses are returned by raindrops and hailstones, so the image maps the regions where these are most intense. Some of the rainbands and the clear intervals between them are also visible on the screen.

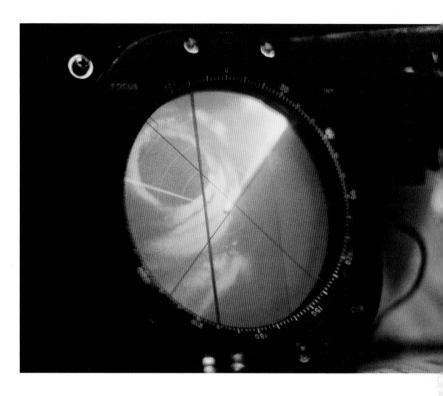

Storm surge, Hurricane Isabel

THE DESTRUCTIVE POWER of a storm surge is shown as it destroys North Carolina State Highway 12. The highway was swept away transforming Cape Hatteras Village into an island. Storm surges are immensely destructive, huge mounds of water raised when tropical cyclones hit the coastline, particularly when the latter is gently shelving. The greatest height ever recorded for a surge was 12.2 m/40 ft at Hatia Island, Bangladesh, on 12 November 1970.

Storm cells over UK, 1987

FALSE-COLOUR SATELLITE IMAGE of the 16 October 1987 storm over the UK. The core of the storm is just right of centre, over the North Sea. Ireland, Wales, and parts of England appear green, as does Brittany and part of France at bottom left. Two massive storm cells over the south coast of the UK span the English Channel. This image was made at 11.00 am, some time after the storm had passed over the UK, uprooting millions of trees and causing extensive damage to buildings in the early hours. Wind gusts of over 160 kph/99.4 mph were recorded.

Extreme erosion due to tropical cyclone

A VISIBLE-LIGHT IMAGE of the Betsiboka estuary, Madagascar, taken from the International Space Station after the passage of Tropical Cyclone Gafilo in March 2004. Extensive deforestation has resulted in massive soil erosion, causing the waters to run red from the laterite soil that has been washed down by the heavy rains. This has led to extensive sedimentation in the river, impeding and preventing the passage of ships.

143

Thunderstorms, Parana Basin, Brazil

THIS IMAGE FROM SPACE illustrates how the anvils from individual thunderstorms may spread to provide a nearly continuous cover of cloud. These clouds are at least 10 km/6.2 miles deep and some show overshooting tops where strong updraughts penetrate the capping inversion and extend a short distance into the stratosphere.

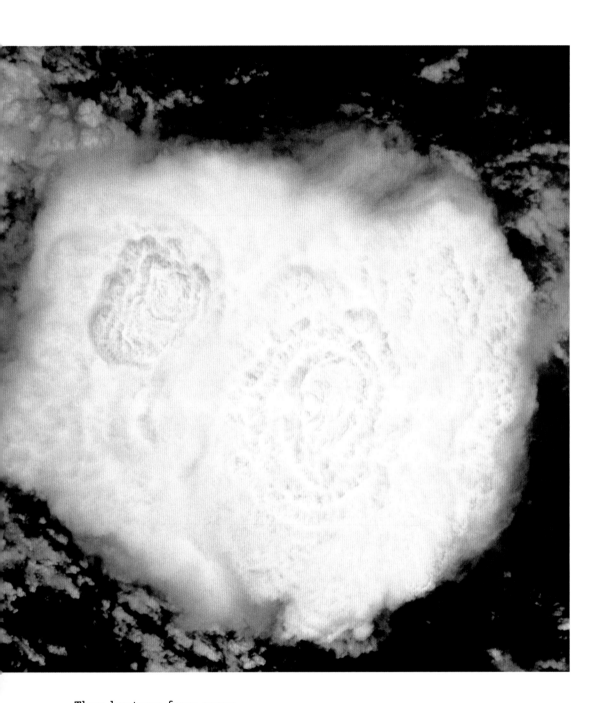

Thunderstorm from space

THIS IS A VIEW of a large cumulonimbus cloud cell seen from an orbiting Space Shuttle. The cell has two major updraughts, the cloud is estimated to be about 50 km/31 miles across and about 13 km/8 miles high. This type and size of cloud is associated with very severe thunderstorms. This photograph was taken by the crew of the Space Shuttle Columbia, launched on 26 April 1993, whilst over the coast of Nigeria.

ALTHOUGH ALMOST EVERYONE HAS SEEN A RAINBOW, few people realise that this is just one of many optical phenomena to be seen in the atmosphere. Indeed, not all rainbows are the same: they actually display different forms and subsidiary features that normally go unnoticed. Fogbows and dewbows are two other related phenomena. Haloes, created by sunlight or moonlight refracted or reflected by ice crystals, are often missed even though it is estimated that at middle latitudes a halo is visible, on average, every three days. Although there are just a few common halo phenomena, there are also dozens of rarer forms. Many are seen only in polar regions, particularly in the Antarctic, where fine ice crystals ('diamond dust') suspended in the air provide ideal conditions for the creation of halo effects.

Some phenomena, such as coronae and iridescence, are commonly unseen because they occur close to the Sun and are lost in its glare, but become visible if the Sun is hidden. Coronae, in particular, are more frequently seen around the Moon, and in fact there are lunar forms of most optical phenomena, with lunar rainbows and haloes being quite common.

Certain optical effects have become more common with the advent of widespread air travel. Glories for example – coloured rings seen on cloud or mist around the shadow of an observer's head – were once seen only by mountaineers, but are now frequently visible around the shadow of the aircraft an observer is travelling in.

The same applies to subsuns – elongated brilliant patches of light, caused by ice crystals – which at one time were only regularly seen in the polar regions. (Glories may be more common nowadays, but an exact mathematical explanation of the mechanism creating them is still unknown.)

The blue of the sky arises through scattering of sunlight by air molecules. The yellow and red skies of sunrise and sunset occur when blue light has been scattered aside, leaving just the longer wavelengths. On very rare occasions, when there have been major volcanic eruptions, particles ejected into the upper atmosphere contribute red light, which mixes with the blue to give a vibrant purple hue to the sky.

One great natural spectacle is provided by the aurora, which occurs high in the atmosphere, between altitudes of 100 and 1000 km/62–621 miles. Although displays are generally seen at high latitudes, in what are known as the auroral zones, roughly centred on the Earth's magnetic poles, on rare occasions displays extend as far as the magnetic equator. Occasionally, aurorae may thus be seen from any place on Earth. Some displays are quiescent but many exhibit an astounding variety of shapes, colours, and motions. Although aurorae may exert a minor, and indirect, influence on weather through their heating of the upper atmosphere, the magnetic storms that accompany major events can damage satellites, disrupt radio communications, and cause electrical surges and blackouts by affecting power lines at ground level.

6 | ATMOSPHERIC OPTICS

THIS HOUSE IN WOOLSTHORPE is the birthplace of Sir Isaac Newton who first gave an accurate explanation of the colours in a rainbow. The most common rainbow is the inner (primary) bow, but the fainter outer (secondary) bow, with a reversed colour sequence, is not rare. The centre of the bows is always the 'antisolar point', the point on the sky directly opposite to the Sun, relative to the observer. The primary bow arises through a single reflection from the back of falling raindrops, while the secondary bow undergoes two reflections within the same raindrops. 'Alexander's dark band' is an area between the bows where light is directed away from the observer.

Lunar corona seen through thin cirriform clouds

CORONAE OCCUR AROUND BOTH THE SUN AND THE MOON, but because of the glare from the Sun, lunar coronae are usually much easier to see. This image shows the inner aureole, with a brownish ring on the outer edge. Additional coloured rings occur farther away from the centre, and these are the ones generally seen around the Sun. The rings arise from diffraction of sunlight by particles in the cloud.

Lunar corona created by pollen

CLOUD PARTICLES ARE NOT UNIQUE in causing coronae. Here, a beautifully coloured corona has been created by an extremely high concentration of pollen grains, released by pine trees in spring in Finland. It is possible, although not confirmed, that the slightly elliptical shape of the corona has been created by the specific shape of the pollen grains. Such coronae are quite common in heavily forested areas, particularly in the northern climatic zone known as the taiga with its vast coniferous forests. Pollen from birch, the hardiest of the deciduous trees, may also create the same effect.

Coronal rings and iridescence in cirrocumulus clouds

BECAUSE OF THE BRILLIANCE OF THE SUN, the inner corona (the aureole) is not seen here, but the outer rings are particularly strong and have violet on the inside and red on the outside. The similar iridescent colours seen in the cirrocumulus billows towards the bottom of the picture demonstrate that both arise from the same mechanism: the diffraction of sunlight by evenly sized particles within the cloud. The more regular the sizes of the particles, the greater the purity of the colours, and the smaller the particles, the larger the radius of the corona's rings.

Iridescence in rocket trail

THE WATER VAPOUR CONTAINED within the exhaust from a research rocket, fired from a distant range, has frozen into ice in the upper atmosphere. The evenly sized particles have created the brilliant iridescence. Such trails bear some resemblance to the natural iridescent nacreous clouds that occur at altitudes of 20–25 km/12.4–15.5 miles, and the far higher noctilucent clouds at about 80 km/49.7 miles. The distorted shape of the trail is caused by variations in the wind speed and direction at different altitudes in the upper atmosphere.

151

ATMOSPHERIC OPTICS

Complex solar halo display

NUMEROUS DIFFERENT RINGS, ARCS AND POINTS OF LIGHT may be seen around the Sun and the Moon. These are caused by the refraction of light through hexagonal ice crystals in various orientations. This picture shows the most common halo effect – a 22° halo – and a larger, fainter, 46° halo. Above each halo is another effect, known as an upper arc of contact, convex towards the Sun. Running through the position of the Sun is a portion of the white parhelic circle, a halo phenomenon that lies parallel to the horizon (appearing curved here because of the camera's optics). Lying on this circle, just outside the 22° halo, are two parhelia (also known as 'sun dogs'), with tails extending outwards along the parhelic circle.

Circumzenithal arc

THIS HALO EFFECT is quite common and usually shows extremely brilliant colours. It takes the form of part of a circle centred on the zenith (the point directly above the observer), and arises when sunlight is refracted through hexagonal prisms of ice. Only the circumhorizontal arc, a halo phenomenon that consists of a brilliantly coloured arc lying parallel to the horizon, which cannot be seen from high latitudes, rivals its purity and strength of colours.

155

Brilliant parhelion (sun dog)

FREQUENTLY, PARHELIA DO NOT APPEAR as simple circular patches of light resembling the Sun or Moon, but as bands of brilliant spectral colours with white 'tails'. In this instance, taken in Alaska in the month of November, the Sun is well outside the frame to the right, and close to the horizon. Although sometimes known as 'winter rainbows' in northern regions, the effect has no relation to water-droplet rainbows, which always appear on the opposite side of the sky to the Sun.

Sun and lower tangent arc

THIS PHENOMENON was photographed from Antarctica. The tiny crystals (known as diamond dust) in the Antarctic air are ideal for the formation of many halo arcs. Here the strongly curved lower tangent arc appears close to the horizon. Despite the glare from the Sun, a slight indication of the parhelic circle (see page 154), which runs through it, may also be seen.

FOGBOWS, LIKE RAINBOWS are always centred on the antisolar point (directly opposite the Sun in the sky). Their radius is about 42°, the same as the primary rainbow. Although the mechanism is the same as that forming rainbows, in this case the droplets are so small that diffraction broadens the bands of colour so that they overlap and appear white. There may be a hint of blue on the inner edge and red on the outer.

Primary rainbow

160

THE MOST COMMONLY SEEN RAINBOW is the primary bow, a circular arc, approximately 42° in radius, which is centred on the antisolar point (the point on the sky directly opposite the Sun). Red appears on the outer edge of all primary bows, with violet on the inside. This bow shows indications of supernumerary bows within the primary bow. The darkness of the sky outside the primary bow is particularly well shown, this area being known as Alexander's dark band.

Rainbow: primary, secondary and supernumerary bows

THIS PHOTOGRAPH, TAKEN AT TAMWORTH IN AUSTRALIA, clearly shows both primary and secondary bows, Alexander's dark band (the area between them), and the pale supernumerary bows inside the primary bow. Secondary bows are generally weaker than primary bows, have a radius of approximately 51°, and show a reversed sequence of colours, with red on the inside. Supernumerary bows lie inside the primary bow and arise through interference between light that has taken slightly different paths through the raindrops.

Crepuscular rays at sunset

NAMED AFTER THE FACT that they appear around sunrise and sunset, crepuscular rays are the shadows of distant clouds or mountain peaks that are cast on to the atmosphere. They appear to radiate from the (hidden) position of the Sun. In this instance, the strong pink to purple coloration of the sky has been caused by fine dust suspended in the atmosphere.

Anticrepuscular rays at sunrise

CREPUSCULAR RAYS MAY SOMETIMES be seen on the opposite side of the sky to the Sun, when they are known as anticrepuscular rays. They appear to converge on the antisolar point, and on rare occasions may be traced right across the sky from one horizon to the other. Here two strong rays converge in the west as the Sun was rising in the east.

163

Previous pages | Stratocumulus and crespuscular rays

A SECOND FORM OF CREPUSCULAR RAYS arise when sunlight strikes through gaps in the cloud cover. Scattering by dust particles or the water vapour in humid air causes the rays of light to be seen. This common effect is known by many popular terms, such as 'the Sun drawing water' or 'Jacob's ladder'.

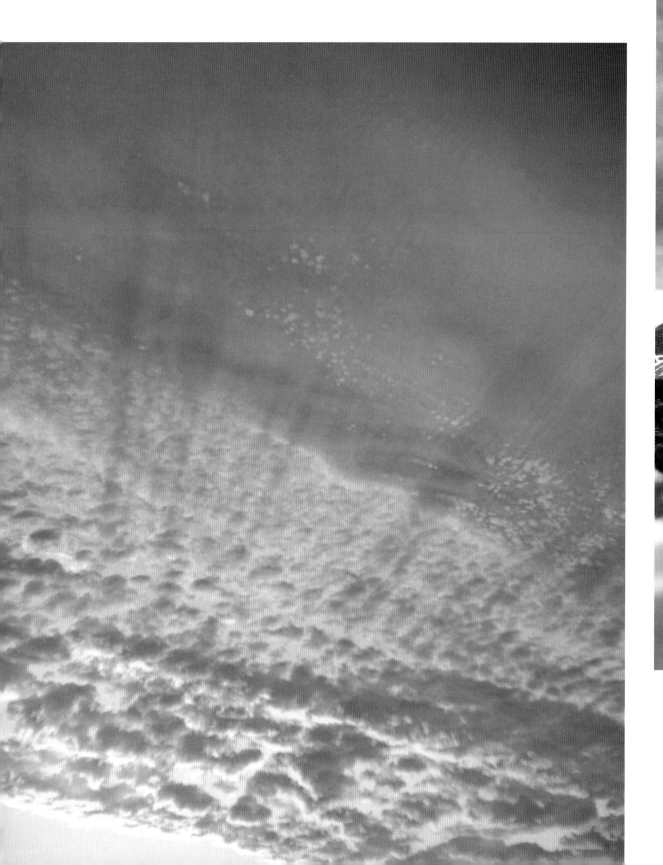

Crespuscular rays

THIS SHOWS STRATOCUMULUS AT SUNSET, with magnificent crepuscular rays. This photograph clearly shows how the shadows of individual cloudlets may be cast on the underside of a layer of clouds, even though the Sun has yet to sink below the horizon.

Alpine glow, Rocky Mountains, British Columbia

THE MOUNTAIN PEAKS, including Mt Assiniboine (3,618 m/11,870 ft high, shrouded in cloud) and Sunburst Peak (left) are bathed in light from the setting Sun. The sequence of striking colours from yellow to purple, called alpenglüh or alpine glow, occurs both at sunset and sunrise, when the Sun is still above the horizon. If the purple light is present, with the Sun below the horizon, the colour bathing the peaks is known as the afterglow.

167

Previous pages | Sun pillar, Arizona

THIS OPTICAL EFFECT is created by light being reflected from the top or bottom faces of flat hexagonal ice crystals floating in the air. The pillars, which may extend 20 or more arc-minutes above and below the Sun, are colourless when high in the sky, but take on the colour of the rising or setting Sun when it is near the horizon. Generally most clearly seen at sunrise or sunset, when they are not overwhelmed by glare from the Sun. Similar moon pillars are seen at night.

Inferior mirage over desert

THIS PARTICULAR MIRAGE is in the Egyptian Western Desert. This form of mirage occurs when the layer of air above the ground is greatly heated. Rays of light are strongly curved upwards, so that an image of the sky or distant objects appears below (hence 'inferior') its true position. The sky appears like water on the surface and, as in this case, multiple, direct and inverted images of distant hills or mountains may be seen.

Superior mirage over sea

THIS MIRAGE IS OVER THE SEA. Superior mirages arise when the surface layer of air is cooler than overlying layers. Rays of light from distant objects curve downwards, in this case producing an inverted image of a distant shoreline, hanging above the peaks themselves. Mirage effect may also cause distant objects to appear compressed or elongated, or be elevated or depressed below the normal horizon.

Fata Morgana mirage

A FORM OF SUPERIOR MIRAGE, where even the surface of the sea or ice floes (or in this case an island) seemingly appears as a vertical, or even overhanging, wall in the distance. Another optical distortion (astigmatism) sometimes comes into play, creating apparent buildings, arches and windows. The name derives from Morgan le Fe, the half-sister of the legendary King Arthur, who was able to create such illusory castles in the air.

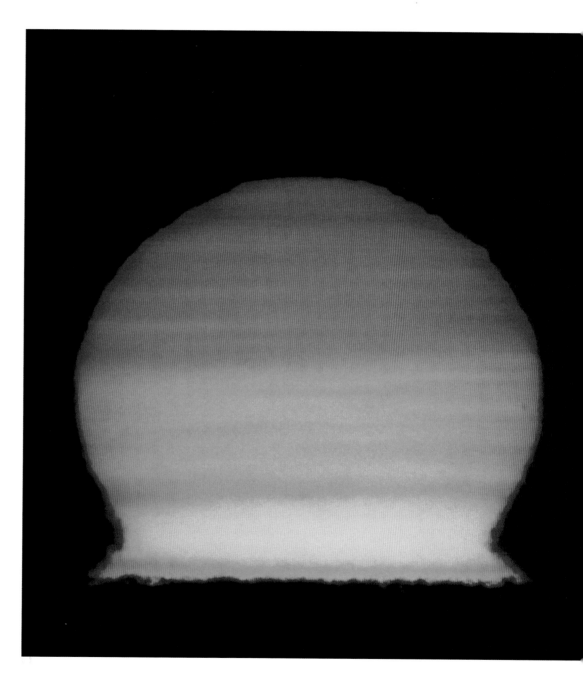

Distorted Sun seen at sunset

THE RISING AND SETTING SUN frequently appears distorted by refraction in layers of air at different temperatures (and hence density). In this case the lower limb of the Sun is actually an inverted image created by a mirage. This type of distortion is quite common and is sometimes called the omega shape, after the Greek letter omega (Ω).

Auroral arc

A QUIET DISPLAY OF THE AURORA BOREALIS. The aurora borealis (northern lights) and the southern counterpart, the aurora australis, are created when highly energetic charged particles, which originally arise in the Sun, cascade into the upper atmosphere, causing the atoms and molecules to become excited and emit light. In this case, the form of aurora is known as a homogeneous arc – quite frequently seen early in a display – although suggestions of a rayed structure are beginning to be evident, a sign that the display is becoming more active.

Auroral corona

THE CHARGED PARTICLES from the Sun penetrate the outer regions of the Earth's magnetic-field region (the magnetosphere) and accumulate in the central plane of a magnetic 'tail' that extends out away from the Sun. When the Sun is particularly active, the particles are accelerated up the tail, towards the Earth and, following the magnetic-field lines, crash into the upper atmosphere in zones around the magnetic poles. An auroral corona appears when the observer is looking directly along the magnetic-field lines.

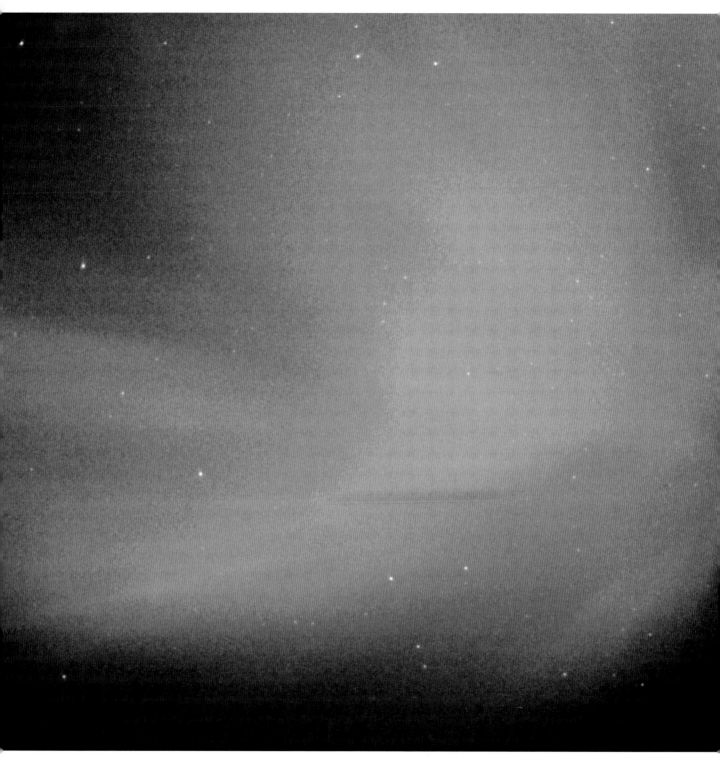

Red aurora borealis display

THE COLOURS SEEN DURING AURORAL displays depend upon the energy of the charged particles that enter the upper atmosphere and the atoms or molecules that are excited. Red displays, such as this one, arise from the excitation of oxygen at high altitudes. A slightly different colour is produced when highly energetic particles penetrate to lower altitudes and excite nitrogen molecules. In the past, such displays were often taken to be distant towns or cities on fire.

Auroral rayed band

THIS STRIKING AURORAL RAYED BAND was photographed from Finland. An auroral band appears as a ribbon of light across the sky. This particular band displays a rayed structure, indicating the lines of the magnetic field. The colour at the bottom of the display arises from oxygen, normally appearing green, rather than the yellowish tint seen here. The violet coloration at the top of the rayed band arises from ionized nitrogen molecules that are not only excited by the charged particles entering the atmosphere, but which also lie in sunlight.

'Auroral green' band

DURING AN ACTIVE DISPLAY, the aurora often develops striking bands. These curtain-like folds frequently show rapid motion, reinforcing the illusion of gigantic curtains waving in the sky. Sometimes several, roughly parallel, bands are visible simultaneously. In this display, photographed over Manitoba in Canada, the green coloration is the highly characteristic 'auroral green', created by a so-called 'forbidden' oxygen emission, i.e. one that only arises when the density of the gas is extremely low, as it is in the upper atmosphere.

Multiple auroral bands

ALTHOUGH MANY AURORAL DISPLAYS are quiescent, sometimes hardly changing for long periods, some displays are highly active, changing shape and position very rapidly. Some, such as this display, over Fairbanks, Alaska, become highly complex with multiple bands, where separate sheets of charged particles are entering the upper atmosphere. In general, auroral displays occur at altitude of between 100 and 1000 km/62.4 and 621.4 miles.

Rayed band

AURORAL DISPLAYS TEND TO START QUIETLY, with irregular patches of light, which may then develop into an evenly glowing homogeneous arc reaching across the sky. Many events do not develop further, but active displays may go on to exhibit a rayed arc, a band (or bands) and rayed bands. Generally the most active displays show rapid motion, often with sudden changes of brightness.

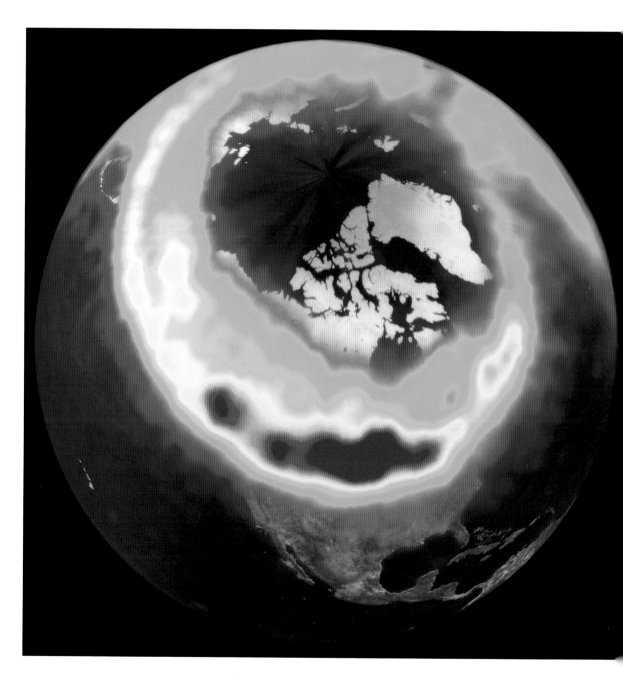

Auroral oval

THIS IMAGE FROM THE POLAR SATELLITE shows an almost complete auroral oval – the region where charged particles are entering the atmosphere at any one time. The aurora is most active in the area that is a reddish brown, and weaker (yellow, blue) elsewhere. The greatest activity occurs shortly after midnight, but some displays are so strong that auroral light may be detected by satellites even around midday, when the display is invisible from the ground.

Aurora australis from space

THE AURORA AUSTRALIS, photographed from the Space Shuttle. During an auroral event, similar displays occur simultaneously at both ends of the magnetic lines of force that link the northern and southern hemispheres. The Shuttle's low orbital altitude (less than 500 km/310.7 miles) frequently carries it through the centre of auroral displays.

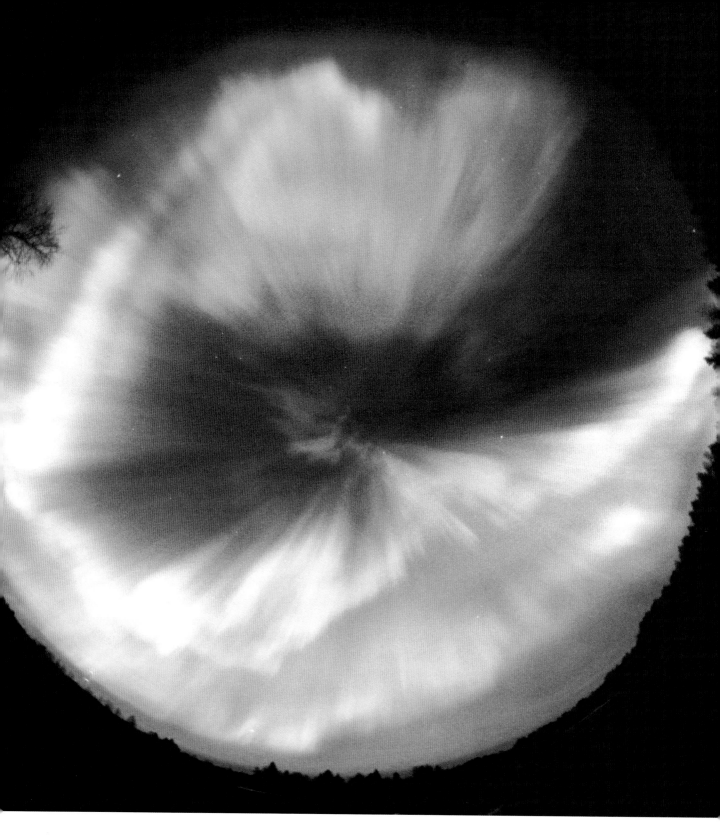

All-sky image: aurora borealis

FISH-EYE LENSES are often used to obtain regular images of the whole sky to monitor auroral activity. Auroral activity commonly occurs in the auroral zone, a region approximately 15–30° from each magnetic pole, where aurorae are more frequently seen. During major solar storms the zone expands, and may even reach as far as the magnetic equator.

Glory

GLORIES RESEMBLE CORONAE, and exhibit similar coloured rings. They are, however, seen around the antisolar point rather than around the Sun or Moon. The phenomenon is caused by the diffraction of light by the water droplets in cloud, fog or mist, although a precise theoretical explanation for the observed radii and colours is still missing. When the cloud or mist is nearby – in this case, just below the photographer on a mountain in Snowdonia – a glory will be seen surrounding the shadow of the observer's own head, but not those of any companions.

184

Icebergs at sunset

INNUMERABLE ICEBERGS ARE SILHOUETTED AGAINST THE SUNSET in this photograph from the
Argentine Islands off Graham Land in the Antarctic Peninsula. The sky shows traces of the rare
purple light, which arises when red light, scattered from aerosols ejected high into the atmosphere by a recent volcanic eruption, mixes with the normal blue light from the sky.

185

EVERY HUMAN ACTIVITY IS DIRECTLY OR INDIRECTLY AFFECTED BY THE WEATHER, so weather forecasting has come to play a vital part – often unrecognized – in modern life. Despite occasional appearances to the contrary, forecasts have become more accurate in recent years, and are gradually being extended to longer periods in the future. However, there are basic, inescapable limits upon our knowledge of the atmosphere and our ability to predict climate changes, so the weather will always spring unexpected surprises and, unfortunately, give rise to unpredictable disasters.

Because weather systems are continually developing and moving, even a single day's forecast requires details of the conditions over a wide area. To forecast the weather just three or four days ahead requires accurate knowledge of the current situation over the whole globe. Vast streams of data flow around the world between the member states of the World Meteorological Organisation (WMO), which is responsible for co-ordinating and operating the global World Weather Watch (the original – and to meteorologists, the most important – 'WWW'). This consists of three interdependent core elements: the Global Observing System, the Global Data-Processing System and the Global Telecommunications System.

The Global Observing System receives standardized observations from thousands of individual observing stations: manned and automatic sites on land; ships, anchored and drifting buoys at sea; aircraft and balloon-borne instrumentation in the atmosphere; and both geostationary and polar-orbiting satellites. The Global Telecommunications Systems transfers this flood of raw data across the world to the meteorological data-

7 | THE OUTLOOK

processing centres (the Global Data-Processing System), where some of the most powerful supercomputers in the world and skilled meteorologists produce local and regional forecasts and severe weather warnings. Some of these may be very specific, such as forecasts of wave-height, for example, the probable paths of tropical cyclones, or the likely effects of volcanic ash plumes on aviation routes. Even longer-term forecasts, such as the strength and timing of El Niño and similar events are increasingly being prepared, thanks to our gradually increasing knowledge of the mechanisms at work in the atmosphere.

The corrected data and forecasts flow back across the Telecommunications System to regional and local centres where they are used by individual meteorological offices and forecasters. The free availability and rapid dissemination of data from anywhere in the world is one of the major examples of successful international co-operation.

In recent decades, orbiting satellites have come to play a major part in monitoring the Earth's weather. From the early satellites that merely transmitted images of weather systems, modern geostationary and polar-orbiting satellites are now able to determine a wide range of atmospheric parameters. These include temperatures at various levels in the atmosphere and at the surface; wind strength and direction (again, at various levels); wave height and direction; atmospheric transparency; the presence of various pollutants; and even surface pressure. Unlike the uneven distribution of ground stations, particularly the sparse surface coverage over the oceans, satellites offer a truly global coverage, day and night, throughout the year.

Imaging the weather: the old...

THIS IS THE VERY FIRST MOSAIC of global weather systems ever obtained, using data returned by the experimental Tiros 9, polar-orbiting satellite, showing the situation on 13 February 1965. Despite its poor resolution by modern standards, it heralded a revolution in the way meteorologists studied weather and climate.

Previous pages | Weather around the world

THIS MOSAIC WAS CREATED from numerous satellite observations showing the weather around the world. The colours and shading approximate those that would be seen by an observer in space. The land is dominated by the green areas of vegetation, and the brown tints of the arid lands in the tropics. The clouds were observed over a period of two days.

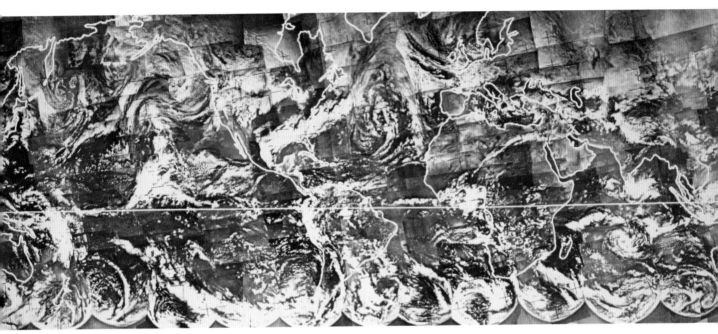

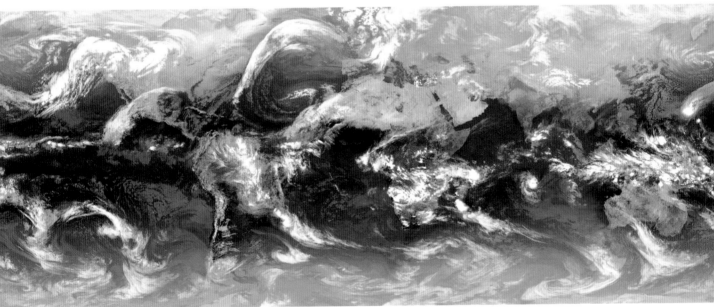

...and the new

A MOSAIC OF GLOBAL WEATHER systems produced from data returned from five geostationary satellites. With continuous coverage of the whole world, such images are now available every 30 minutes – in this case for 15:00 GMT on 18 January 2005. Because half of the globe is in darkness at any time, observations in the infrared were used as the principal source of data, and the colours were computer-generated to approximate those that would be seen by the human eye.

The Earth's cloud cover in 3-D

THIS COMPOSITE IMAGE OF THE EARTH showing surface temperature (various colours) and three-dimensional (3-D) cloud cover (white), was created by integrating multiple satellite data, and is the first representation of the Earth's cloud cover in three dimensions. Temperature is colour-coded, from red (high) through to dark blue (low). A strong El Niño event in progress in the Pacific Ocean is visible as a red band in equatorial regions. Height of the clouds is greatly exaggerated.

Meteosat images of the Earth

THESE THREE IMAGES OF THE EARTH, all obtained on 15 January 1989, illustrate how imaging the world at different wavelengths provides valuable information to meteorologists. The visible-light image (below) was obtained in a region of the spectrum that corresponds quite closely to the human visual range. A notable feature is the band of cloud clusters and thunderstorms along the Equator, marking the position of the Intertropical Convergence Zone (ITCZ). Here trade winds from each side of the Equator converge and there is strong upwelling as part of the global circulation. The two circulation cells on each side of the Equator are known as the Hadley cells. Clear skies occur over the sub-tropical high-pressure zones (particularly over North Africa and Arabia), where the descending limbs of the Hadley cells lie. Here the air warms with descent, preventing the formation of cloud.

THE REGION OF THE SPECTRAL CHANNEL used for this image (left) was specifically chosen to reveal the water-vapour content of the atmosphere, with black showing low humidity and white high humidity. The humid, upwelling air along the Intertropical Convergence Zone is prominent as are swirls of alternating dry and humid air associated with low-pressure areas in both northern and southern hemispheres. The latest Meteosat Second Generation (MSG) satellites record images in twelve spectral channels, providing highly detailed information about conditions at various levels in the atmosphere.

IN THIS INFRARED IMAGE (below) the hottest regions are black, and the coldest pure white. The cold cloud-tops along the Intertropical Convergence Zone are prominent, and the dark areas show the high temperatures prevailing over the Sahara, Arabia, and the Middle East. Shades of grey provide an indication of variations in sea-surface temperatures. Long trails of cloud in both hemispheres indicate the approximate location of the Polar Front, where cold air from the poles encounters warm air flowing out of the sub-tropical high-pressure zones.

191

Geostationary satellite image: Africa and Europe

GEOSTATIONARY SATELLITES ARE STATIONED 35,900 km/22,308 miles above the Equator, where their orbital period is precisely equal to 24 hours, so they remain essentially stationary above one point on the Earth (because the Earth's gravitational field is not uniform, they do drift slowly out of position, and this is corrected by the use of thrusters.) This Meteosat image is centred on longitude 0° (the Greenwich Meridian) and latitude 0°. Satellites observe the Earth at various regions of the spectrum, each of which gives a black-and-white image. Channels may be combined to give images with approximately correct colours.

Geostationary satellite image: North and South America

THIS FALSE-COLOUR IMAGE WAS OBTAINED from the GOES-E satellite, located over the Equator at longitude 75°W. Apart from Meteosat at longitude 0°, three other geostationary satellites provide complete coverage around the world: GOES-W at 135°W, GOMS at 76°E, and GMS at 140°E. Although coverage is excellent in tropical and sub-tropical regions, the curvature of the Earth introduces limitations at high temperate latitudes and over the poles. Monitoring of these regions is carried out by polar-orbiting satellites at a much lower altitude.

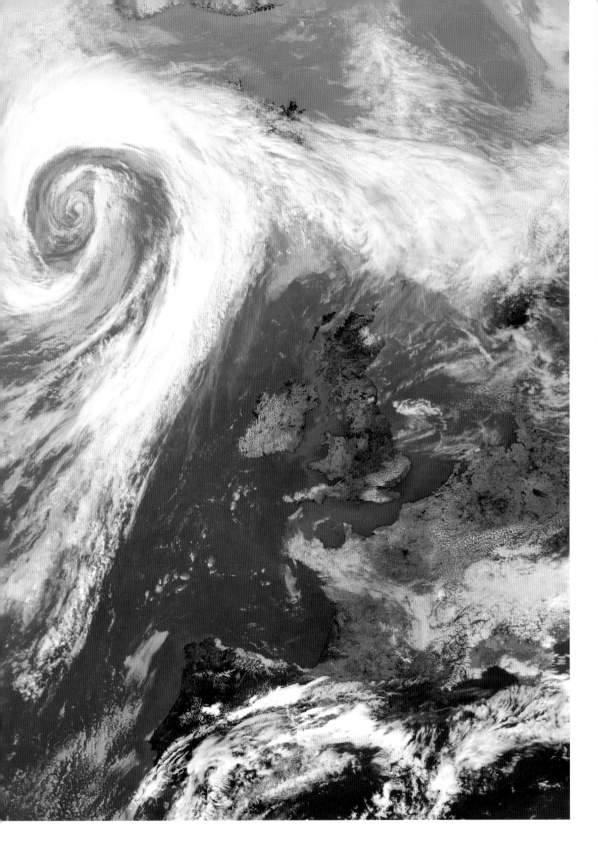
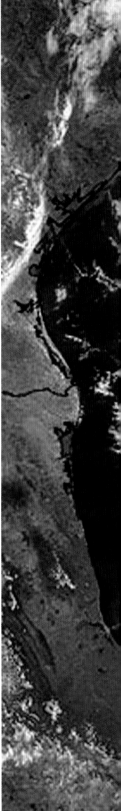

Western Europe from a polar-orbiting satellite

194 POLAR-ORBITERS PASS AT MUCH LOWER ALTITUDES (800–1000 km/497–621 miles) than geostationary satellites, and have highly inclined orbits to cover the polar regions. As the Earth rotates
beneath the orbit, successive passes cover adjacent (overlapping) swathes of the surface, with
two overhead passes of a particular region every day. In this image of western Europe an anticyclone (a high-pressure area, with cloudless skies) lies over the British Isles. The swirl of clouds
marks a depression (low-pressure area) over the Atlantic.

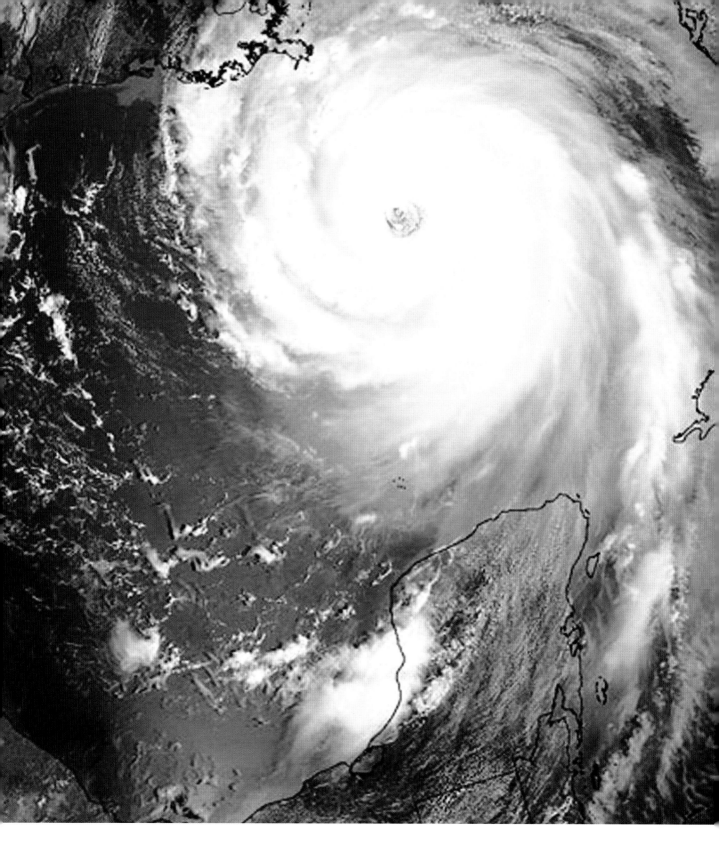

Tracking hurricanes: Hurricane Katrina

SATELLITES PROVIDE VITAL INFORMATION in monitoring the position of tropical cyclones (known as hurricanes over the Atlantic). Here Hurricane Katrina, with its vast swirl of clouds and central eye is seen on 28 September 2005, just before it hit Lousiana and Mississippi, devastating the Gulf seaboard and inundating New Orleans.

True-colour image of Ireland

THIS IMAGE WAS TAKEN BY THE TERRA SATELLITE, which carries instrumentation specifically designed to monitor various factors that may be involved in climate change. The rugged coast-line of western Ireland, with its distinctive rias (valleys flooded by sea-level rise following the last Ice Age) and mountainous areas appear brown, with grasslands and peat bogs, green. At top and centre right, snow covers the mountains of Scotland and North Wales, respectively.

An extratropical low-pressure system

PHOTOGRAPHS, SUCH AS THIS IMAGE OF A DEPRESSION, taken during a Space Shuttle mission, are not normally used for forecasting, but they provide useful information about the growth and decay of weather systems. This depression formed over the Gulf Stream near Cape Hatteras in North Carolina. Most of the spiral of cloud consists of an occluded front, where the cold front (bottom, nearest the camera) has caught up with the warm front (short section at extreme top right), and lifted the warm air away from the surface.

197

The smoke plume from Mt Etna

VOLCANIC PLUMES POSE A GREAT HAZARD for aircraft. Mt Etna (3,323 m/ 10,902 ft) in Sicily, is Europe's largest and most active volcano: an orange lava flow is visible running south from the crater. On a clear day, such as here, when this image was obtained by the SPOT Earth-resources satellite, the plume is easy to track. When there is heavy cloud cover and complex weather patterns, determining the probable danger zone becomes far more difficult.

Atmospheric water vapour

THIS MAP FROM THE AQUA SATELLITE shows areas of low humidity around North and South America as light blue, and those with high humidity, such as over the Caribbean, as dark blue. In the north, snow and ice appear yellow, but the small yellow patches in the tropics indicate individual thunderstorms with heavy precipitation. The dark stripe (right) shows loss of data over the Atlantic.

199

Water vapour over the world's oceans

AS WITH THE PRECEDING IMAGE, AREAS OF LOW HUMIDITY appear light blue, and those with high humidity (mainly in the tropics) are dark blue. Ice and snow over both the Arctic and Antarctic appear in yellowish tints, and some desert areas (particularly the Atacama Desert in South America, and the deserts in Central Asia) have a slight yellow-green tinge. As the name implies, the Aqua satellite is specifically designed to monitor the Earth's hydrological cycle.

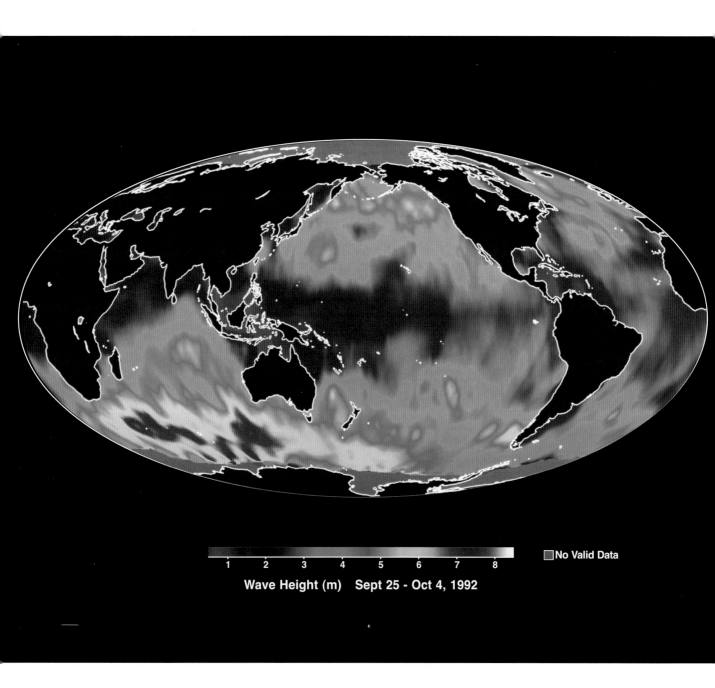

No Valid Data

Wave Height (m)   Sept 25 - Oct 4, 1992

Global map: mean wave heights

THE TOPEX/POSEIDON SATELLITE, orbiting 1,336 km/830 miles above the Earth, carried a radar altimeter to determine mean wave height across the globe, colour-coded here from magenta (less than 1 metre), through blue, green, and yellow to red (7–8 m/23–26 ft). The data was obtained at the end of September and the beginning of October, 1992. The largest waves are in the south Indian Ocean and the Southern Ocean near Antarctica, and are typical of the season (southern winter/spring).

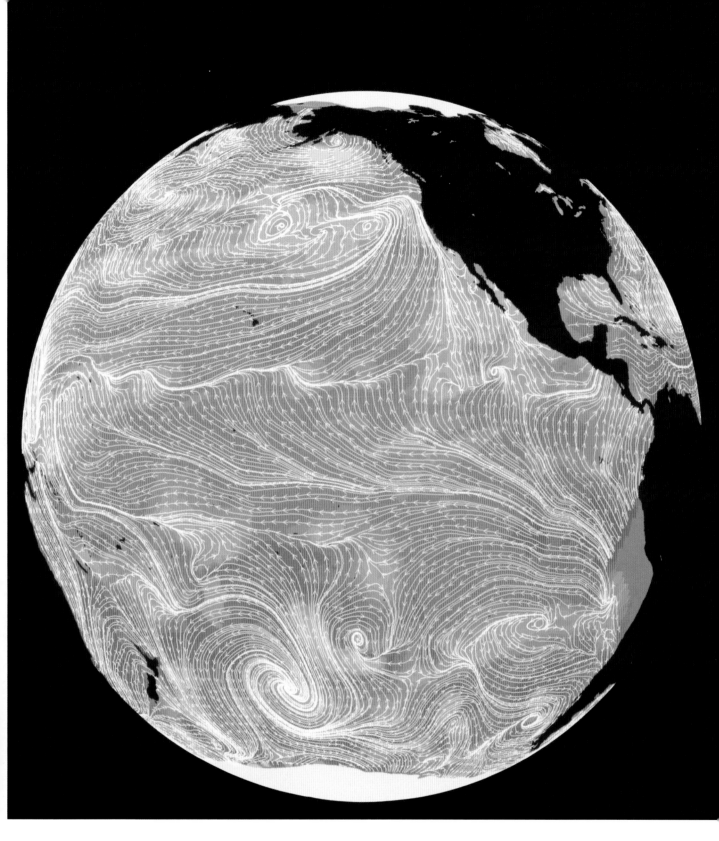

Pacific-Ocean wind speed and direction

THE ARROWS REPRESENT WIND DIRECTION and the colours wind speed over the Pacific Ocean. Blue shows wind speeds of 0–14 kph/0–9 mph, purple and pink speeds of 15–43 kph/10–27 mph, red and orange speeds of 44–72 kph/28–45 mph. High wind speeds occur near a depression located in the South Pacific (where the circulation is particularly marked) and also south of Alaska. The distinct, narrow band with low wind speeds beneath the Intertropical Convergence Zone is visible, running right across the Pacific just north of the Equator.

Source of the Gulf Stream

SEA SURFACE TEMPERATURES AROUND THE FLORIDA PENINSULA are shown colour coded, from grey and dark blue (0–2°C/0–35.6°F) through light blue, green, yellow and orange to red (22–26°C/72–79°F). Large-scale eddies are visible in the Gulf of Mexico at bottom left. The very warm Florida Current, passing between Florida and the Bahamas, is joined by the very broad, warm Antilles Current, flowing east of the Bahamas, to become the Gulf Stream. This powerful current flows north along the eastern seaboard of North America, before turning east off Cape Hatteras to become the North Atlantic Current.

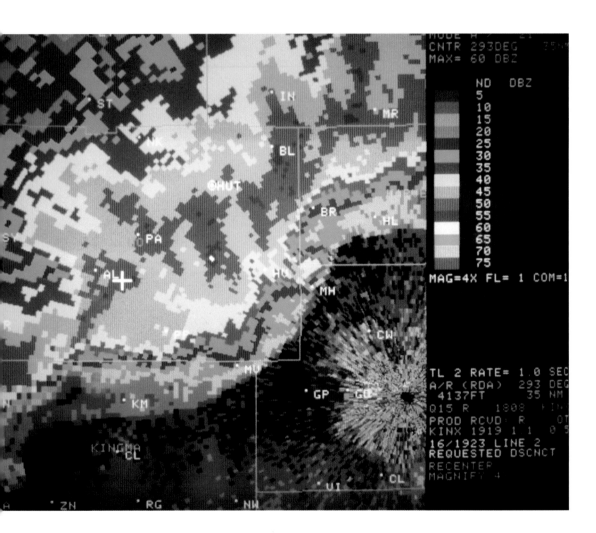

Radar image of a squall line

GROUND-BASED RADAR PROVIDES INFORMATION about the intensity of rainfall and is often used in 'nowcasting' (the preparation of short-term forecasts for no more than 6 hours ahead). The most advanced form (known as Doppler radar) also provides information about the direction and strength of the wind. Such details are vital in preparing warnings of imminent severe weather such as extreme hailstorms or tornadoes.

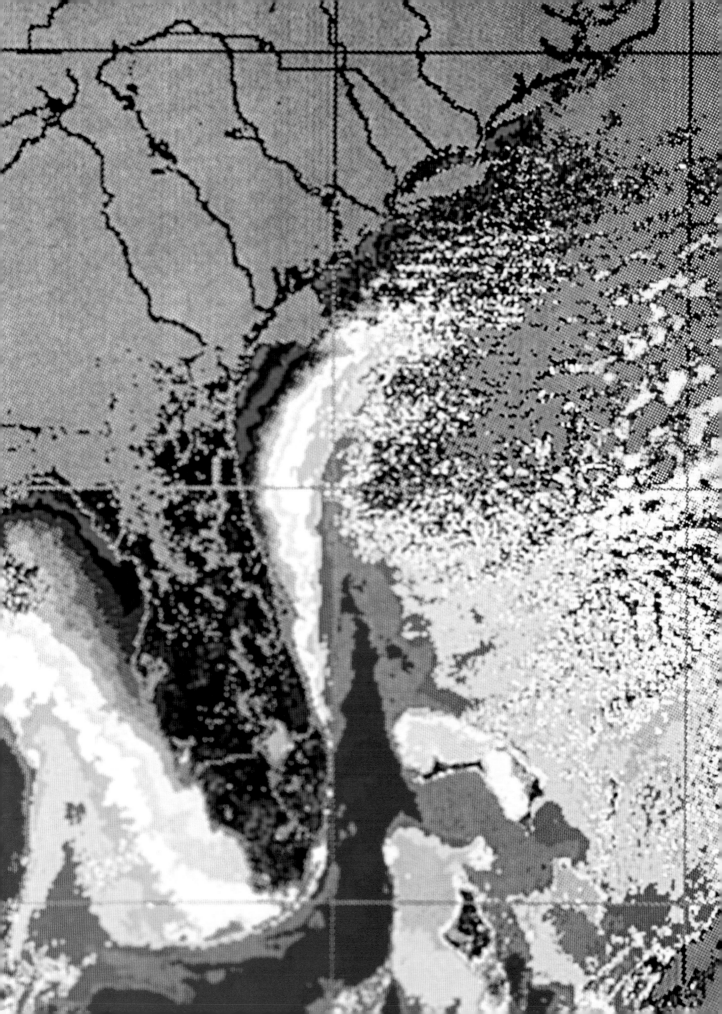

Launching a weather balloon

DESPITE THE TREMENDOUS ADVANCES in measuring various parameters of the atmosphere from orbit, balloon flights with relatively simple instruments that radio data to the ground still play a vital part in weather forecasting. This type of balloon (known as a 'Jimsphere', after its inventor Jim Scoggins) is designed to be exceptionally stable in flight. It is slightly pressurized and the four hundred conical projections increase its stability. Hundreds of balloons are released daily from sites all over the world, here from the European Space Agency's launch site at Kourou in French Guiana.

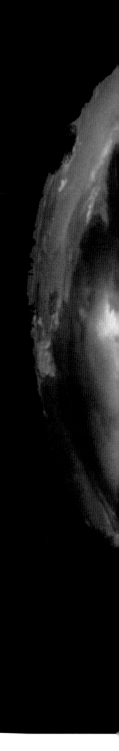

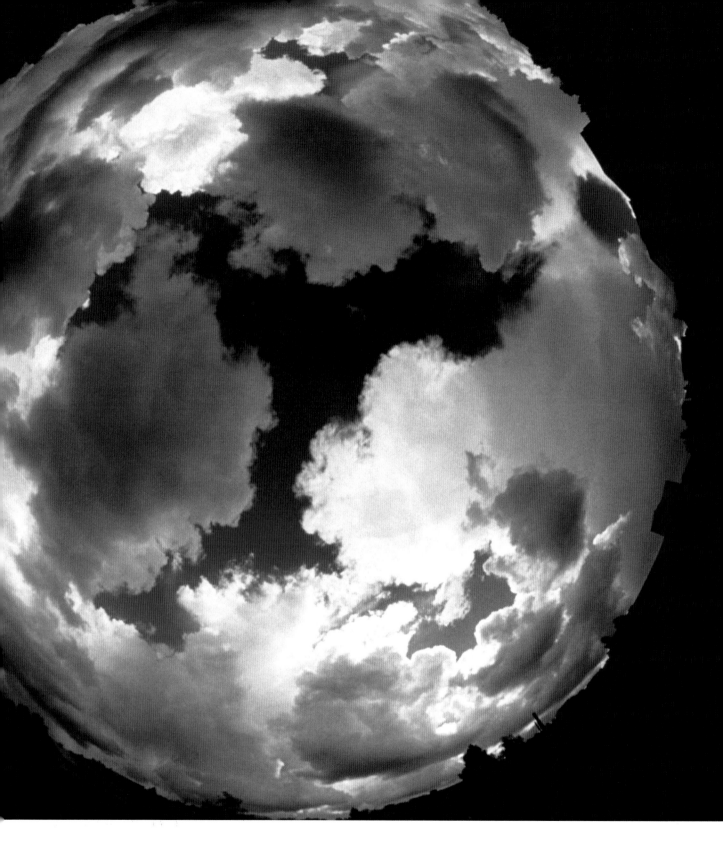

Cloud cover assessment

A VIEW OF THE WHOLE SKY OBTAINED with a fisheye lens, assessing the amount of cloud cover (in this case low-level cumulus clouds). Specially-designed orthographic projection lenses such as this one are used to obtain an image of the whole sky. With such lenses the percentage of the image covered by cloud is precisely related to the percentage of the actual sky covered by cloud. This information is then used directly in weather reports, and in the preparation of forecasts.

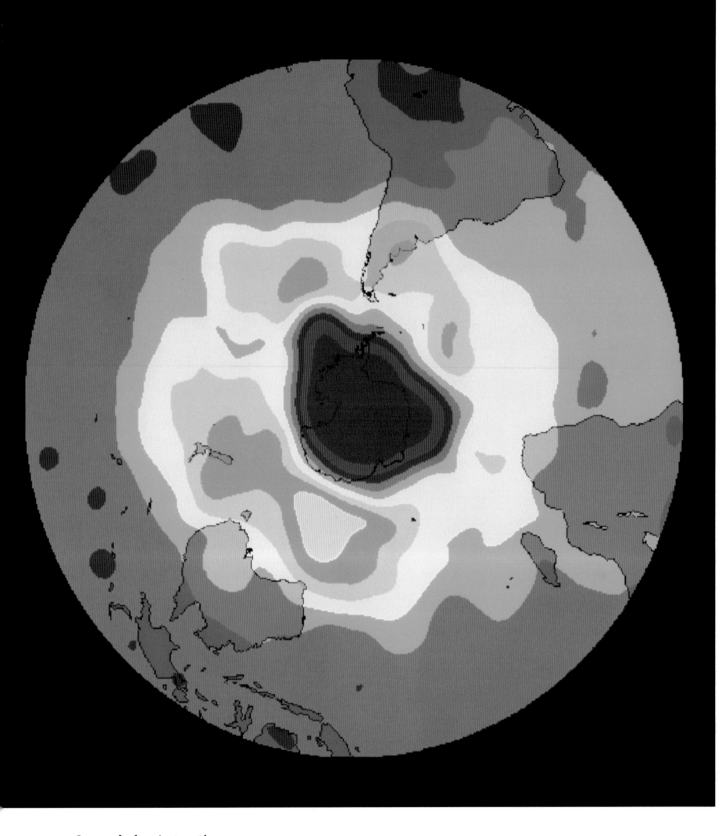

Ozone hole: Antarctic

THE GREATEST DESTRUCTION OF OZONE occurs in spring (in this case, October), when strong
sunlight returns to the polar region after mid-winter darkness. The Antarctic ozone hole is particularly strong because a polar vortex of high-speed winds surrounds the pole, flowing essentially
without obstruction right round the world. This cuts the polar region off from the general global
circulation, preventing the replenishment of ozone from farther north.

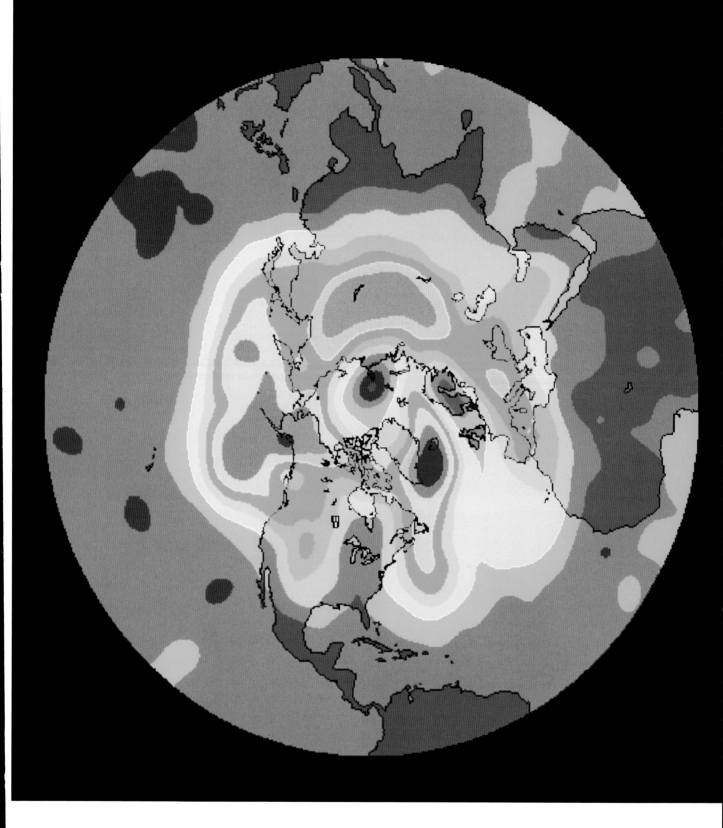

Ozone hole: Arctic

IN THE ARCTIC, THE OZONE HOLE is much smaller than its Antarctic counterpart. This is primarily because the northern polar vortex is much weaker than the vortex in the south. The northern land-masses (and particularly mountain ranges such as the Rockies in North America) disrupt the circulation, and greater mixing occurs between polar air and that from lower latitudes.

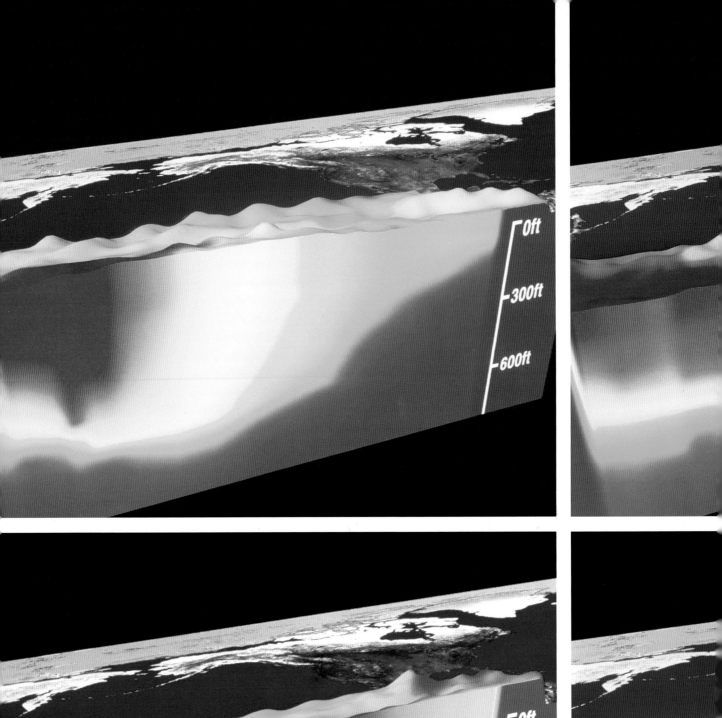

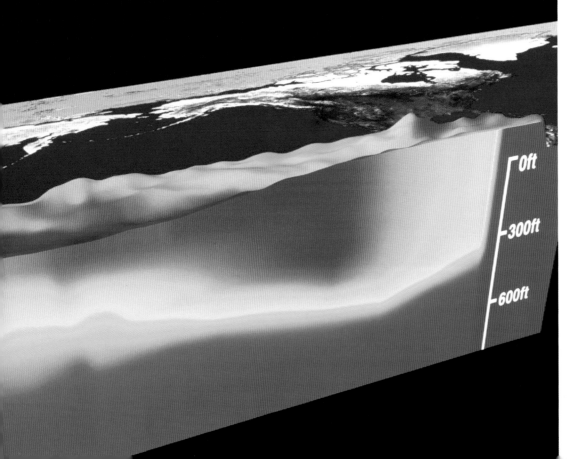

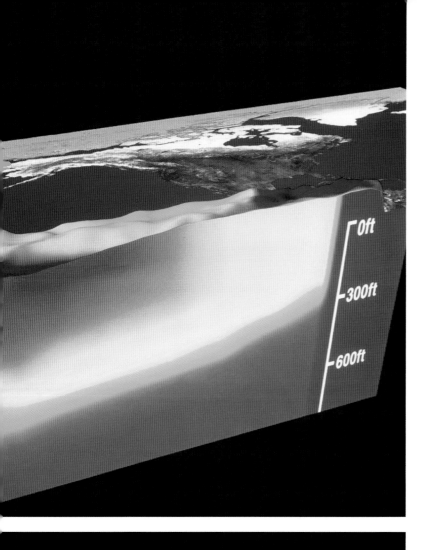

Development of El Niño event

COMPUTER REPRESENTATIONS of four stages of an El Niño event, showing conditions right across the Pacific. The map at top gives an idea of the relationship of the section to the Americas (top right) and Asia (top left). Sea temperatures vary from red (30°C/86°F) to dark blue (8°C/46.4°F). The height of the sea surface is shown in greatly exaggerated relief.

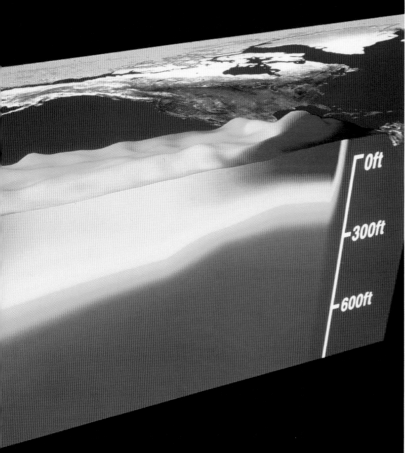

IN THE FIRST IMAGE (top left) TAKEN IN JANUARY 1997, there is a deep pool of warm water – and its associated rainfall – in the western Pacific (i.e., on the Australian and Asian side), where the ocean surface is actually higher than near the coast of South America. Near the latter there is upwelling of cold water, leading to highly productive fishing grounds. The thermocline (the division between warmed surface water and cold deep oceanic water) slopes steeply towards the west. Such conditions correspond to the 'normal' (non-El Niño) state.

IN THE SECOND IMAGE (top right) FROM JUNE 1997, the surface winds driving the warm water to the west have weakened, and the pool of warm water is spreading eastwards across the Pacific. Upwelling along the coast of South America has ceased, and a very thin, warm surface layer has arisen. The slope of the thermocline has decreased dramatically.

IN THE THIRD IMAGE (bottom left), OBTAINED IN NOVEMBER 1997, the El Niño event is well under way. The deep pool of warm water and the accompanying area of heavy rainfall have nearly reached South America. The ocean surface is now higher in the east than in the west. Drier conditions (and an Australian drought) prevail in the western Pacific. The fishing industry off the coast of Peru has collapsed.

BY MARCH 1998 (bottom right), the pool of warm water has reached the South-American coast. The associated, unaccustomed, torrential rain caused flash flooding, mudslides, and loss of life in Peru. Other events, farther afield, particularly in the northern Pacific and in the Indian Ocean, are also linked to El Niño events.

211

Global sea temperature, July

IT IS PROXIMITY TO THE OCEAN and the temperature of its surface waters that largely determine the weather and climate occurring in any particular region. Ocean currents transport more heat from the tropics towards the poles than the atmospheric circulation. This computer model of sea surface temperatures is based on satellite data. The sea temperature varies from 35°C/95°F (yellow) in the tropics, through red, blue, purple and green to -2°C/28.4°F (black) in the polar regions. Land appears grey.

Global sea temperature, December

THE YELLOW TINT INDICATING the warmest sea surface temperature has shifted south in this computer-generated image derived from satellite data for December (during the southern hemisphere's summer). It is striking, however, that some regions of the North Atlantic off North Africa are now warmer than they were in July, where water warmed during the northern summer has spread across the ocean.

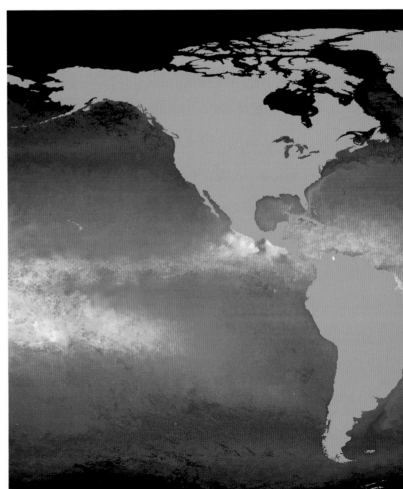

212

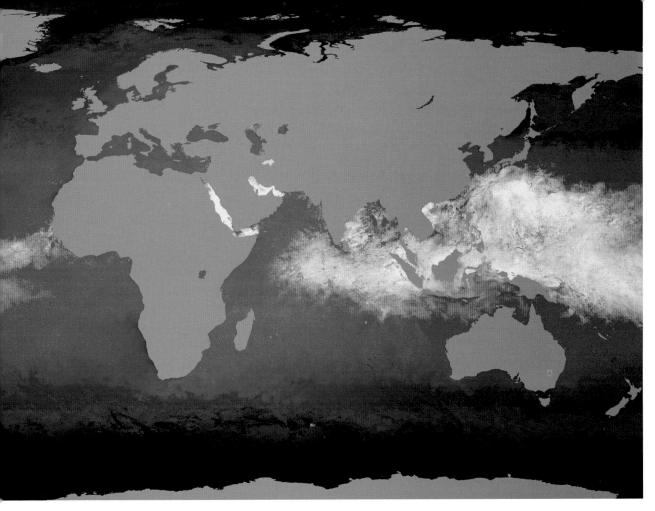

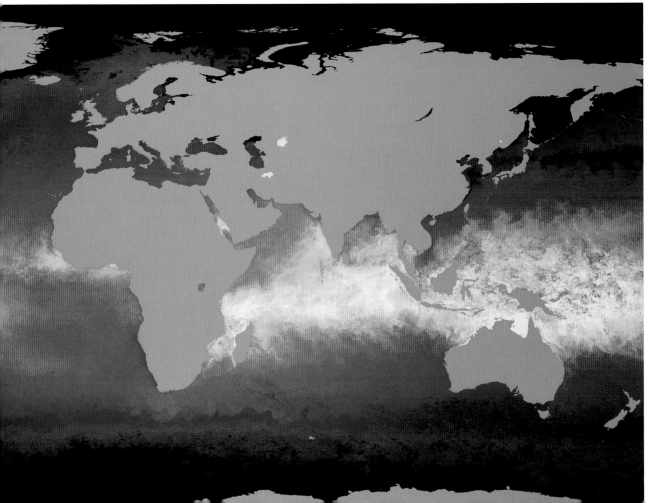

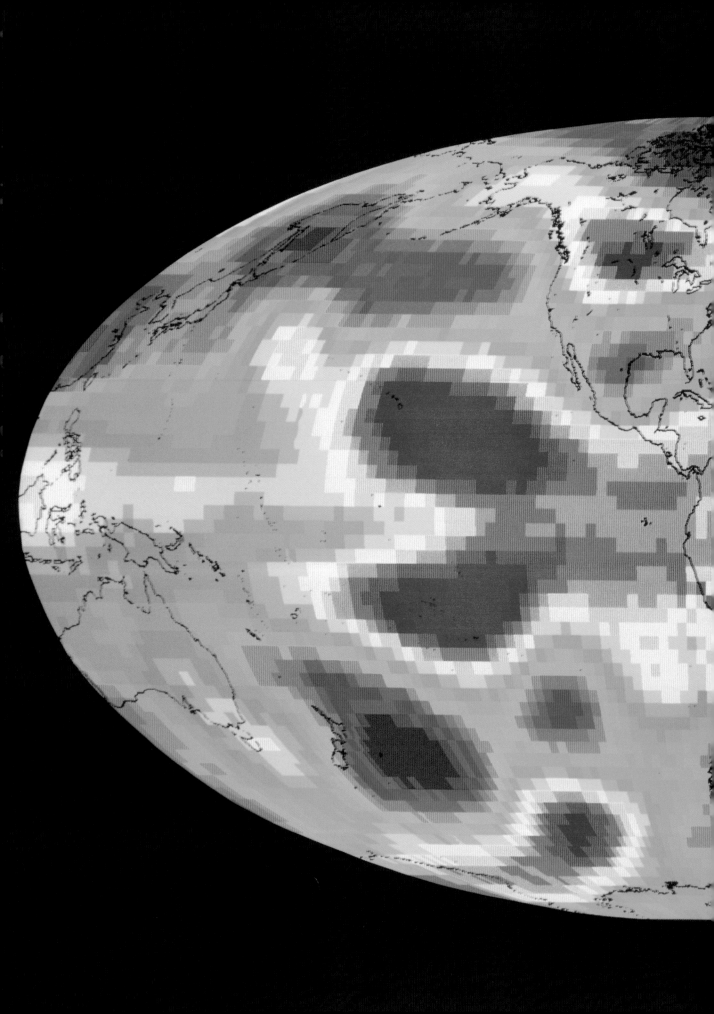

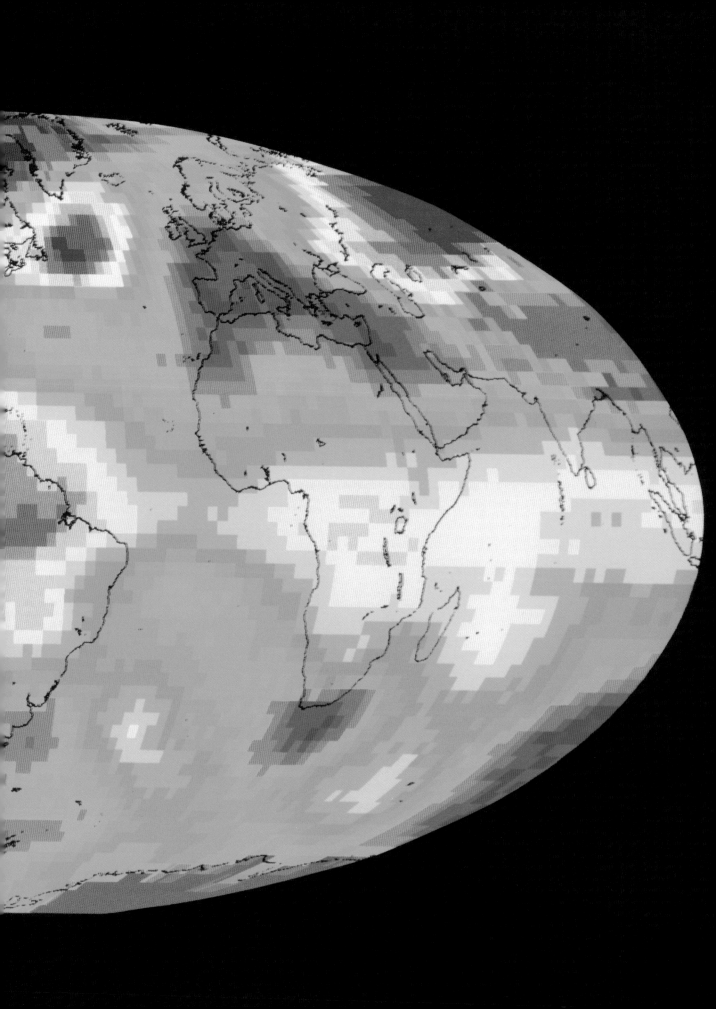

ANOMALIES ARE REGIONS where one of the atmosphere's properties (in this case temperature) varies significantly from the long-term average. Here, red areas indicate regions where the temperature is higher than the long-term average, and blue areas where it is lower. The red area over the Pacific, off the coast of South America, is typical of the pattern shown during El Niño events. This climatic phenomenon recurs roughly every ten years, causing major changes in ocean circulation, as described earlier, and severely affecting climate and biological productivity. Associated effects (known as teleconnections) influence the weather over most of the world, being strongest over the southern hemisphere and the northern Pacific.

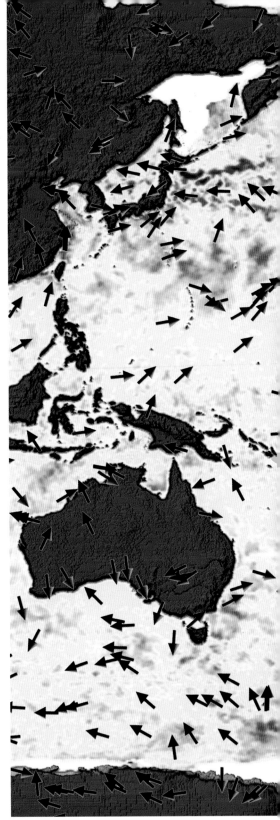

Butterfly effect and chaos theory

FOLLOWING A DISCOVERY MADE while modelling weather systems on a computer, meteorologist Edward Lorentz published his paper, 'Predictability: Does the Flap of a Butterfly's Wings in Brazil Set off a Tornado in Texas?'. The idea, related to chaos theory, states that small events can make a massive difference further down the line, and has been popularized as 'the Butterfly effect'. Lorentz's paper, however, showed that ultimately the question was inherently unanswerable. Errors in weather forecasting are inevitable because of inadequate global coverage, errors in the initial data, our uncertain knowledge of the physics involved, and unavoidable approximations in the equations used for computation. The Butterfly effect implies limits on predictability, not that small events necessarily have major consequences. The photograph shows a blue morpho butterfly of a type that inhabits the rainforests of northern South America.

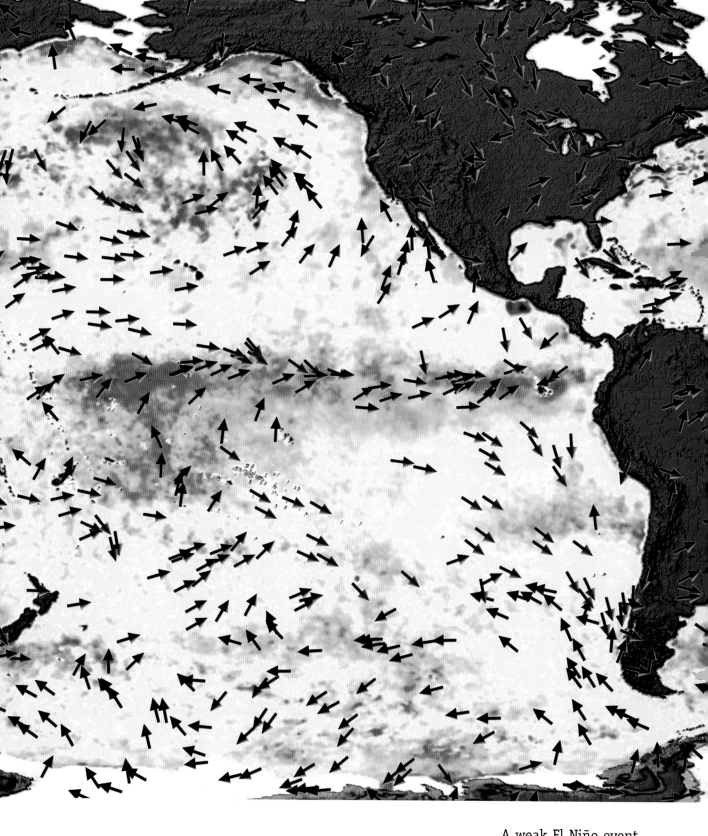

A weak El Niño event

SEA-SURFACE TEMPERATURES and wind directions in March 2003 from satellite data. Orange areas are hotter than normal and blue areas cooler. The tropical Pacific is now monitored continuously by satellites and the nearly 70 moored buoys in the Tropical Atmosphere-Ocean Array (TAO). The El Niño phenomenon is part of a larger system, known as the El Niño Southern Oscillation (ENSO), discovered in 1923, which links atmospheric pressure at Tahiti, in the

central Pacific, to that over the Indian Ocean, as measured at Darwin in northern Australia. When one is high, the other is low. ENSO events are now fairly well understood and it is possible to predict how they will evolve, and affect conditions off South America and outside the immediate area. A strong El Niño brings heavy rains to southern California. The event shown here was weak, because the pool of warm water lay in mid-Pacific, not close to South America.

217

North Atlantic depression

A FALSE-COLOUR IMAGE OF A FRONTAL SYSTEM over the North Atlantic Ocean. The United Kingdom and France may be seen at bottom right, and Greenland's ice cap at top centre. The swirling lines of clouds mark the fronts associated with the depression, with the lowest pressure at the centre of the spiral. The warm front has just covered Ireland, with partially broken cloud cover between it and the cold front to the north. The cold and warm fronts merge to form an occluded front that runs into the centre. Isolated showers dot the cold air sweeping in behind the cold front. Low-level clouds are shown as yellow, high level clouds as white.

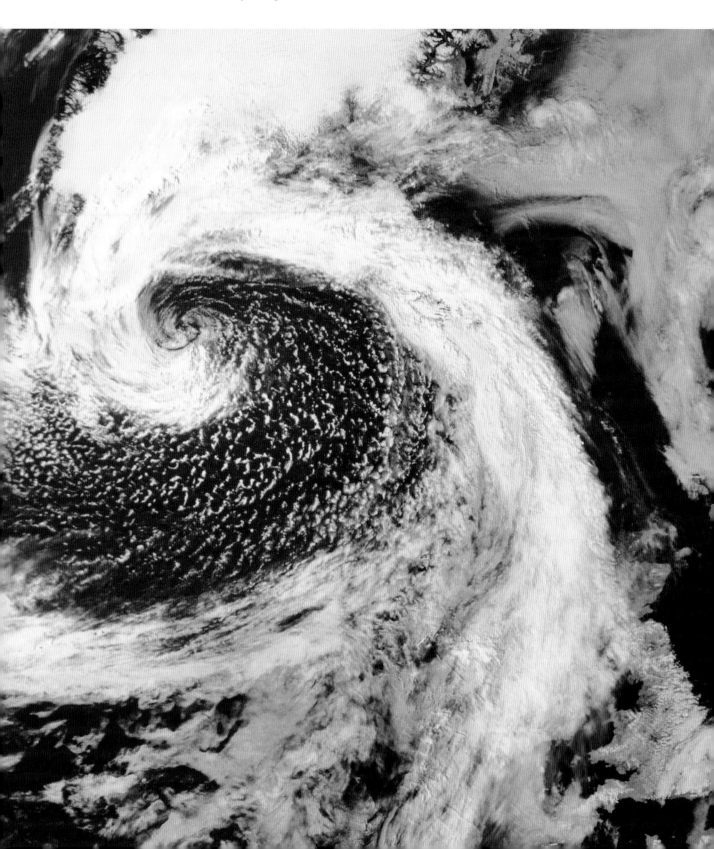

Southern-hemisphere depression

THIS SATELLITE IMAGE SHOWS A DEPRESSION off the southern coast of Australia. The Bass Strait is visible, but Tasmania is beneath the cloud. Winds circulate around low-pressure centres in opposite directions in the two hemispheres, anticlockwise in the north, and clockwise in the south. Unlike the image of the northern-hemisphere depression, the warm front (bottom) is indistinct, and the main feature is the long, clearly defined, cold front that runs on into the occluded front that is tightly wrapped around the centre.

THE EARTH DISPLAYS AN ASTOUNDING VARIETY OF CLIMATE ZONES, from the icy polar deserts to the humid rain-forests of the tropics. The climate at any location is influenced by many factors, the most important of which are the latitude; the proximity to the ocean and the sea surface temperature found there; the direction and strength of the prevailing wind; and the altitude.

The hot, humid air rising over the equatorial regions flows north and south at altitude, and descends at the middle latitudes, producing the mid-latitude high-pressure zones, where clear skies and high temperatures prevail. It is here that the world's major hot deserts are located, such as the Sahara and Arabian deserts in the northern hemisphere, and the interior desert of Australia in the southern. Because of the Earth's distribution of land masses, most of the high-pressure zones in the southern hemisphere lie over the oceans, where there are large regions of low precipitation.

Air flows from the mid-latitude highs, some back towards the Equator, giving rise to the trade winds, and some towards the poles. Because of the rotation of the Earth the air is diverted towards the right in the northern hemisphere and towards the left in the southern. This gives rise to two zones of westerly winds. In the south, because land masses are small, high, persistent winds occur in a broad belt above the Southern Ocean, surrounding Antarctica.

In the north, the greater land masses and, in particular the existence of major mountain chains, such as the North-American Rockies and the high Tibetan Plateau, lead to great variability in the location of the Polar Front,

where warm air from the south encounters cold air flowing south from the Arctic. Despite this variability, certain high- and low-pressure areas tend to recur frequently and determine both the short-term weather and the longer-term climate of particular regions. Apart from the mid-latitude highs (such as the Azores–Bermuda High over the Atlantic), high pressure over Siberia, Arctic Canada, and Greenland dominates winter conditions. Less persistent low-pressure areas such as the Aleutian and Icelandic Lows also play a considerable part.

The interiors of large land masses have a continental climate, with low annual precipitation, extremely cold temperatures in winter and high temperatures in summer. Areas close to the ocean enjoy a milder, maritime climate, with greater precipitation, and smaller temperature ranges. This is particularly marked on the western sides of the continents, especially in western Europe and on the west coast of North America, and also on the southern tip of South America.

Great seasonal variations occur in regions subject to monsoon winds, where there is a major shift in the direction of the prevailing wind with the season. In particular, cold air flowing from the Siberian High in winter produces northerly and northeasterly winds and dry conditions. As summer approaches, the high-pressure area collapses and the jet stream moves to the northern side of the Himalaya and the Tibetan Plateau, allowing southwesterly monsoon winds to bring humid air and heavy rains to southern Asia.

8 | CLIMATES OF THE WORLD

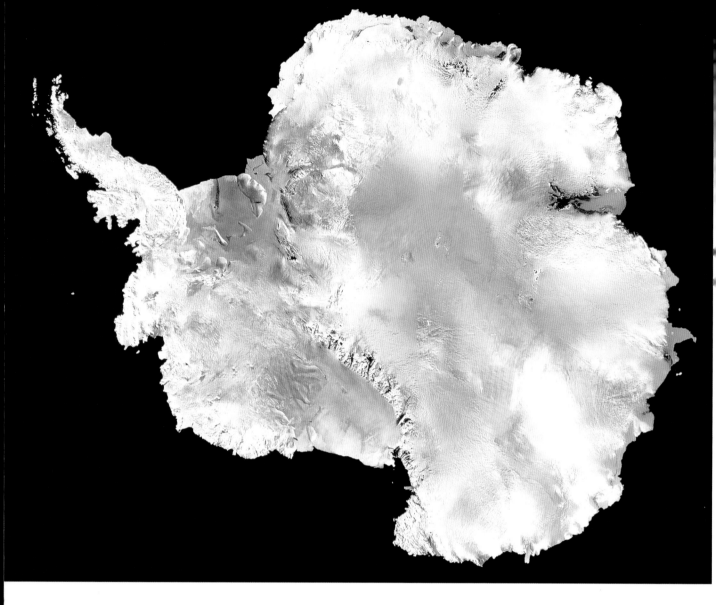

Antarctic ice cap

AN INFRARED IMAGE OF ANTARCTICA shows the continent almost covered with snow and ice. It exerts a major influence on climate in the southern hemisphere and it is divided into two distinct regions by the Transantarctic Mountains: East Antarctica (on the right), an immense high plateau, and West Antarctica (on the left). Antarctica is technically classified a desert, as precipitation in the interior is extremely low. Despite this, the ice in East Antarctica's central plateau reaches an altitude of 4,270 m/14,000 ft, even though bedrock is about sea level. West Antarctica is generally lower, and bedrock over large areas is below sea level. It does, however, contain the highest point of the continent, Vinson Massif, at 5,140 m/16,864 ft.

The McMurdo Dry Valleys

WITH THEIR UNIQUE ECOSYSTEM, the valleys lie in the Transantarctic Mountains, bordering the Ross Ice Shelf, and are separated from the main area of East Antarctica (top) by rock barriers. With their minimal precipitation, they remain free of snow and ice, unlike the valleys (right), where glaciers drain ice from the plateau towards the Ross Ice Shelf. In summer, temperatures are high enough to melt the ice on the lakes in the Dry Valleys.

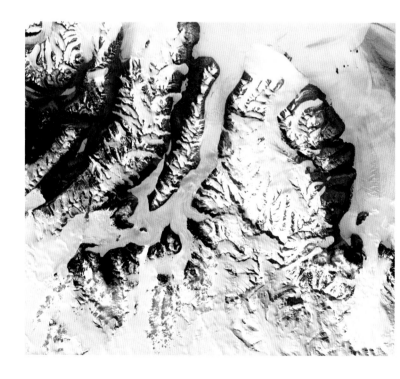

Previous pages | Antarctic moonrise

THE MOON RISES over an Antarctic landscape where a few isolated mountain
peaks (nunataks) rise above the ice cap. Taken at the end of summer, this shows
the moon over Alexander Island, the largest of the islands lying off the Antarctic
Peninsula, .

Edge of Antarctic ice cap

THE IMAGE SHOWS ICEBERGS CALVING INTO THE WEDDELL SEA. Much of the ice
draining from Antarctica through individual glaciers combines with neighbouring
streams to form vast floating ice shelves. Here, icebergs break away from the
Ronne Ice Shelf, the second largest ice shelf (after the Ross Ice Shelf on the opposite side of the continent). The average thickness of the ice here is about 150
m/492 ft.

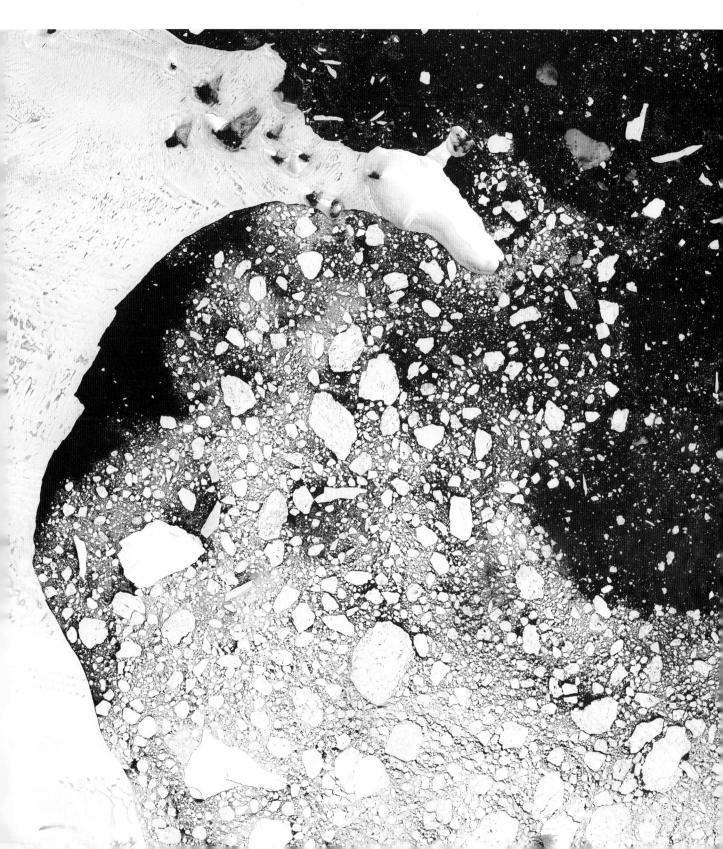

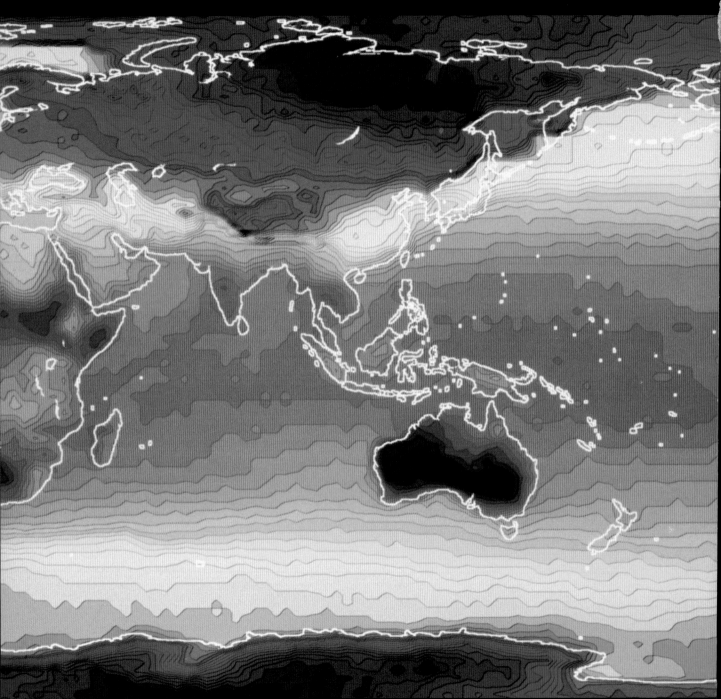

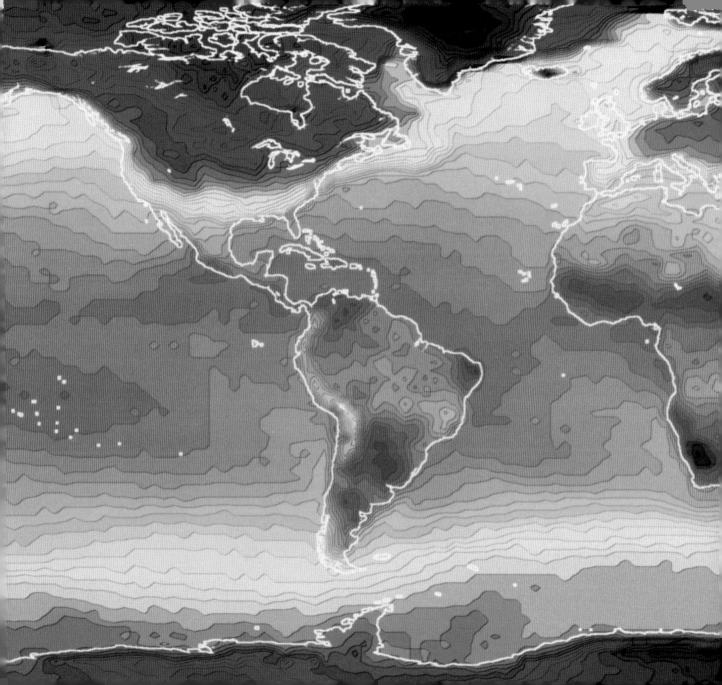

A TYPICAL SATELLITE MAP of the average surface temperature (in this case for January), showing the immense contrast between the tropics and the polar regions. Temperatures are represented as follows: mauve (almost black in the northern hemisphere) = -38°C/-36.4°F or less; blue = -36– -12°C/-32.8–10.4°F; green = -10–0°C/14–32°F; yellow = 2–14°C/35.6–57.2°F; pink and red = 16–34°C/60.8–93.2°F; deep red and black (as in Australia) = 36–40°C/96.8–104°F.

Ice cover in the Arctic

THIS INFRARED IMAGE MAP SHOWS THE MAXIMUM (winter) extent of the Arctic polar ice. This consists of sea ice, unlike the Antarctic ice cap, and includes the whole of the Arctic Ocean, semi-enclosed seas such as the Baltic and the Sea of Okhotsk (top right), and coastal areas of Canada and Alaska. The permanent (summer) ice cap is much smaller, touching land only at the northern edge of Greenland and some of the Canadian islands. A separate, land-based, ice sheet covers most of Greenland, and this has a major effect on the weather and climate of northern Europe.

Barren polar tundra

THE CLIMATIC ZONE KNOWN AS THE TUNDRA is the relatively flat, treeless region of the Arctic that lies between the ice cap and the northern limit of tree growth, where it gives way to the forested taiga. Many areas, such as here in northeastern Greenland, may be completely devoid of any plant life, because of low precipitation, short summers, freezing temperatures and violent winds. Exposed rocks are broken into small fragments by the repeated freeze/thaw cycles. There is no corresponding zone in the southern hemisphere, but similar climatic conditions (known as the alpine tundra) occur over the Tibetan Plateau and in parts of the high Andes.

Edge of Greenland ice sheet

THE GREENLAND ICE SHEET is second only to the Antarctic ice sheet in size, and covers about 1,834,000 sq. km/708,111 sq. miles, roughly 80 per cent of Greenland. Bedrock is near sea level, but the northern dome reaches an altitude of over 3,000 m/9,843 ft, and the southern about 2,500 m/8,202 ft. The ice sheet is largely contained by coastal mountains but gaps in the mountain ranges allow individual glaciers to reach the sea. In winter, the ice sheet becomes the centre of the Greenland High, a high-pressure area from which frigid air streams out over the surrounding regions.

Hummocky ground in the tundra

THE WHOLE OF THE TUNDRA IS TREELESS, but certain areas may support mosses, lichens, and dwarf shrubs during the brief, warm summer season. The hummocks are a feature of much of the tundra, and occur in areas of permafrost (permanently frozen subsoil). They (and the related areas of stones sorted into polygonal patterns, known as patterned ground) are created through the repeated freeze/thaw cycles, when larger stones migrate to the edge of the polygons. Any vegetation tends to hug the ground in the hollows between the hummocks, where it gains some slight shelter from the icy winds. The photograph shows Clavering Island in northeastern Greenland, at about latitude 74°N, where the pack ice melts in summer, as shown by the ice-free inlet in the distance.

Pingo, northern Canada

THIS IMAGE SHOWS A PINGO IN THE PERMAFROST ZONE OF NORTHERN CANADA. Pingoes are highly characteristic features of permafrost regions. They occur in the swampy surface layer above the permafrost, and consist of a lens-shaped core of ice, covered by a thin layer of earth or gravel. They gradually grow over a period of many years and some have become substantial hills, reaching heights of 65 m/213 ft and diameters of 800 m/2,625 ft or more.

Pine forest in the taiga

THE TAIGA IS THE SUB-ARCTIC CLIMATIC ZONE that lies in an immense swathe across the northern hemisphere, between the treeless tundra and the boreal grasslands of Eurasia and the Canadian prairies. Overall, the zone experiences moderate amounts of rain, with extremely severe winters and peak summer temperatures of 15°C/59°F or less. It mainly consists of vast coniferous forests (as here in British Colombia, Canada) with a few very hardy deciduous species, notably birch. There is a distinct treeline on the mountains, above which trees are unable to grow, generally because of the decrease in temperature with altitude.

Canadian prairie

THE CANADIAN PRAIRIES EXPERIENCE A CONTINENTAL, boreal snow climate with moderate annual precipitation, extremely severe winters and cool summers. Only one region south of Edmonton, Alberta. Canada may be considered to have a semi-arid steppe climate. In Eurasia, a similar climatic zone covers a vast tract of land from the border of Poland and Belorus in the west, to the River Yenisey near Krasnoyarsk in Siberia in the east. This photograph of a farmhouse west of Medicine Hat, Alberta emphasises the wide open spaces of these grasslands: it is an insignificant speck in a vast expanse of grass and sky.

Snow-covered Siberian steppes

HERE THE DENSE BOREAL FOREST TO THE NORTH and the sub-arctic taiga to the east give way to the semi-arid steppes lying to the south. The river on the right, running south to north (bottom to top) is the Ob – a major river in Siberia. The roughly circular, dark grey area on the extreme right at a bend in the river, is the city of Barnaul. The gently undulating grassland plain is covered in snow, broken by remarkably straight, tree-filled river valleys, dark green where snow has fallen from the trees. The valleys are hundreds of kilometres long, some draining east into the Ob, but others flowing west into another great Siberian river, the Irtysh, out of the image to the left. Middle-latitude steppes have a continental climate with low annual rainfall and high temperatures during the summer.

Temperate rainforest

THE CLIMAX VEGETATION (NATURALLY-BALANCED PLANT COMMUNITY) IN MARITIME west-coast climate zones (otherwise known as warm temperate rain climates) is a temperate rainforest, seen here in the Olympic National Park, Washington state, USA. This climatic zone generally has rainfall throughout the year, mild winters and moderately cool summers. These conditions are found over a relatively restricted narrow band on the West Coast of North America, but over most of western Europe, where the climax vegetation is often deciduous, rather than coniferous, trees. Similar conditions occur in southern Chile, a small area around Johannesburg in South Africa, southeastern Australia, and New Zealand.

Mediterranean olive groves

THE MEDITERRANEAN CLIMATIC ZONE is a warm temperate one, with generally mild winters and moderate amounts of rain, and a dry season in summer. Around the Mediterranean itself summer temperatures are generally above 22°C (71.6°F), but slightly lower summer temperatures are found in areas with a similar climate in California, South Africa, and southern Australia. These conditions are ideal for the cultivation of olives, vines, and citrus fruit.

235

The Florida Everglades

SIMILAR TEMPERATURE CONDITIONS to areas with a Mediterranean climate are found in the humid sub-tropical climatic zone, but with a greater overall annual precipitation. In large regions of the United States (including the Everglades), Argentina, Brazil, China, southern Japan, and a small area of New South Wales in Australia, rainfall is high throughout the year. Northern India, Burma, and northern Vietnam have less rainfall in winter.

Ancient hominid footprints in the savanna

THIS TRAIL OF FOSSILIZED FOOTPRINTS in volcanic ash, at Laetoli in Tanzania, probably created by *Australopithecus afarensis*, an early hominid, 3.6 million years ago. (Those on the far right were created by Hipparion, an extinct three-toed horse.) Early hominids may have developed fully upright, bipedal gait when climate change occurred in East Africa. Tropical rainforest gave way to humid, open grasslands of the savanna, where upright stance would be advantageous. Savanna regions have high temperatures through the year and usually one major rainy season. They cover swathes of central Africa, South America (particularly Brazil), and some regions of India, Indochina, northern Australia, and Central America.

Tropical rainforest, Ecuador

THE ORIENTE TROPICAL RAINFOREST in eastern Ecuador (left), lying east of the Andean Cordillera, is part of the great Amazonian rainforest. Tropical rainforests, which are primarily found here, in central Africa, and in Indonesia and New Guinea, experience considerable rainfall throughout the year. There is never less than 60 mm/2.4 in per month, although there is often a distinct rainy season, when rainfall may be much higher. The humidity, combined with the relatively constant, high temperatures, means that the vegetation continues to grow throughout the year, leading to an extremely rich and productive ecosystem, with an incredible range of species of plants, insects, birds, and animals.

A cloud forest

CLOUD FOREST VEGETATION (RIGHT) develops at altitude in a few isolated areas in tropical regions. Here, low temperatures due to altitude, and prevailing winds, rising over the mountains, bring almost continuous rain or mist throughout the year. This results in a forest consisting of a canopy of tall trees and dense understorey of shrubs and plants. The conditions are particularly suitable for epiphytic plants such as orchids and bromeliads, which flourish in upper canopies. This cloud forest lies at 2,200 m/7,218 ft on the eastern slopes of the Andes in Ecuador. There are signs that, with warmer temperatures, cloud forests are starting to shrink and that their lower boundaries are starting to retreat to higher altitudes.

Giant groundsel on Mt Kilimanjaro

THE RELATIVELY ISOLATED, EQUATORIAL MOUNTAINS of Africa (Kilimanjaro, Elgon, and Kenya, in particular) have a high-altitude moorland zone, notable for giant groundsel and giant lobelia. These species have adapted to the low night-time temperatures at the high altitudes. (In the groundsel, the rosette of leaves closes around the central bud at night to protect it from the frost.) This photograph was taken on Mt Kilimanjaro, Tanzania, at an elevation of 3,700 m/ 12,139 ft.

Monsoon flooding on the plains of Northern India

MONSOON CLIMATES SHOW A MARKED REVERSAL between winter and summer wind directions, accompanied by a short dry season in winter, and very heavy rains over the rest of the year. The most pronounced monsoon conditions apply over Asia, with particularly heavy summer rainfall in India, Bangladesh, Burma, Thailand and the Philippines. Somewhat similar conditions occur in west Africa (in Sierra Leone and Liberia) and in small areas of South America (Guyana, Surinam, and French Guiana), and northern Australia.

Wind-sculpted rocks

YARDANGS ARE A CHARACTERISTIC FEATURE of desert or semi-desert areas where there is little precipitation, but strong, persistent winds. The soft, poorly consolidated rocks are eroded into ridges and gullies, parallel to the direction of the prevailing winds. The differences in height may be considerable, often reaching several metres. Yardangs are one of the few landforms created solely by the action of the wind, but are found in many areas of the world. This yardang is in the Sahara, in Egypt's Western Desert. The most extensive fields occur in Central Asia, such as around Lop Nor, the dried-up salt-lake bed in the Sinkiang Autonomous Region of China.

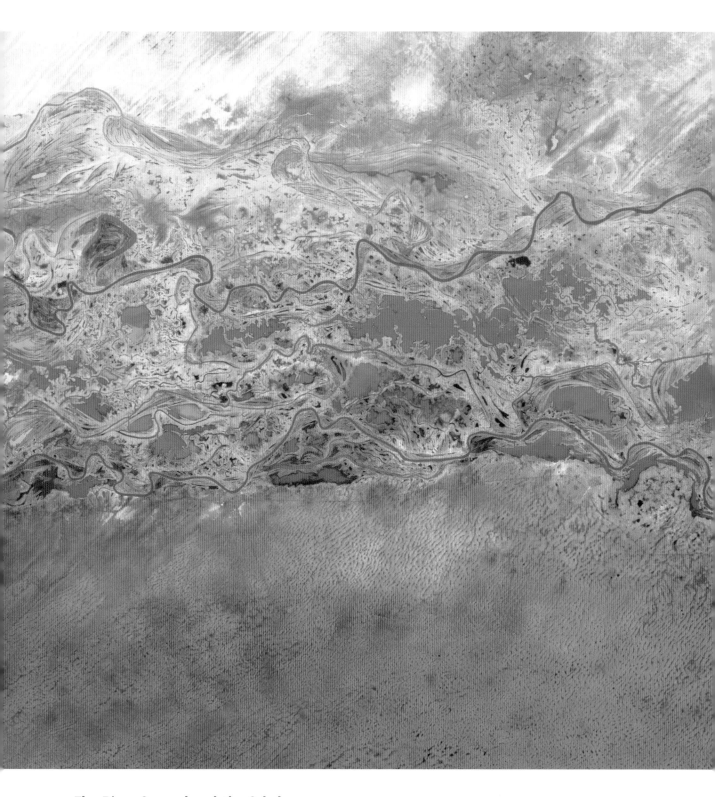

The River Senegal and the Sahel

IN THIS FALSE-COLOUR, SATELLITE IMAGE, the River Senegal in West Africa, with its extensive flood-plain (grey) marks the boundary between the arid desert of Mauretania with its linear sand dunes (top) and the more humid savanna of Senegal (bottom). This boundary region is known as the sahel, and some years ago there was considerable concern that the semi-arid, steppe-like lands of this region were expanding at the expense of the savanna farther south, perhaps as a result of over-grazing and deforestation in the sahel region. It now appears, however, that there is a long-term, natural climatic cycle. The northern and southern boundaries of the sahel move backwards and forwards, and the width of the sahel itself expands and contracts over a period of decades.

Western Desert, Egypt

THIS COMPOSITE OF INFRARED- AND VISIBLE-BAND IMAGES shows part of Egypt's Western Desert, northeast of the Dakhla Oasis, and on the edge of the vast Sahara Desert. Linear sand dunes lie parallel to the direction of the prevailing wind. The areas of exposed rock (brown) are scarred by dry river channels (known as wadis), cut during the rare flash floods that affect the region at long intervals.

Giant sand dunes, Namib Desert

THE DESERT IN NAMIBIA is one of the driest areas in the world. There is very little rainfall, because the Benguela Current, which flows north along the coast, is extremely cold, suppressing evaporation. The little precipitation that occurs is primarily derived from fog, which forms when warm air is cooled by the cold surface waters and then drifts inland. The sand dunes in the Namib Desert are the largest found anywhere in the world, and some reach 350 m/1,148 ft in height.

Extensive wadis in coastal desert of Yemen

IN THE COASTAL DESERT OF YEMEN as in all deserts, the surface rocks are generally subjected to extreme temperature changes from day to night. The expansion
and contraction fractures the rocks, leading to boulder-strewn areas and large
expanses of rocks and gravel as well as the sand that is commonly thought to be
characteristic of deserts. Any rainfall leads to flash floods which rapidly, and
deeply, erode the unvegetated surface and produce the characteristic wadis (dry
river channels) and outwash fans of water-borne debris.

Overleaf | Salar de Uyuni, Bolivian Altiplano

THE ALTIPLANO (ALSO KNOWN AS THE PUNO) is a high-altitude desert in the
Peruvian and Bolivian Andes. It consists of a series of basins, separated by mountainous ridges. In the north lies Lake Titicaca, and to the south Lake Poopó, the
Salar de Coipasa, and the vast Salar de Uyuni. The latter is an enormous salt pan,
the remnants of an ancient lake, about 10,600 sq km/4,093 sq. miles, lying at an
altitude of 3,656 m/11,995 ft. The polygonal pattern is reminiscent of the hummocks and patterned ground found in areas of the tundra or polar desert.

245

NEARLY ALL METEOROLOGISTS AND CLIMATOLOGISTS AGREE THAT GLOBAL WARMING IS TAKING PLACE, primarily through an increasing concentration of carbon dioxide in the atmosphere. It is exceptionally difficult to build suitable models to predict the sequence of events that is likely to occur and the rate of warming. By comparison, predicting the weather a few days ahead is a relatively simple task, because such predictions primarily involve just the atmosphere.

Predicting future climate is far more complex. Many additional factors must be taken into account. The oceanic circulation is exceptionally important in determining climate, because it transports great quantities of heat from the tropics to the poles. One fear is that global warming will cause a substantial portion of Greenland's icecap to melt or calve as icebergs. This would introduce large quantities of fresh water to the North Atlantic where, under current conditions, extensive evaporation causes the water to become highly saline, and thus dense. The water sinks and then flows south along the oceanic floor, establishing a circulation that involves all the world's oceans. In the Southern Ocean this bottom water is reinforced by frigid water sinking in the Weddell Sea and the flow passes into the Indian and Pacific Oceans, eventually rising to the surface and returning via various branches to the North Atlantic. This circulation (known as the Great Ocean Conveyor Belt) is technically called a thermohaline circulation, because the flow is driven by differences in temperature and salinity. If this flow were to be interrupted it would have major consequences for the climate of many areas of the world, most notably northern Europe.

In the atmosphere itself, the role of clouds is extremely important: the greater the cloud cover, the greater the amount of solar energy that is reflected back into space. Yet clouds also act as a blanket, preventing heat from the surface from escaping, so their effects are complex. With increased temperatures there will be increased evaporation and increased cloud cover. A recent discovery is that there is evidence that rising temperatures have led to increased precipitation over East Antarctica, and that this is acting as a temporary buffer against rising sea levels. Eventually, however, the Antarctic ice will begin to flow more rapidly into the sea, and sea levels will inevitably rise. It has also recently been established that meltwater is now penetrating to the base of the Greenland ice sheet, so that in summer large areas are no longer frozen to the bedrock, as had been thought, and are flowing seaward at a faster rate. This is of even more immediate concern.

The pollutants introduced into the atmosphere that produce extensive haze over large parts of the world also appear to have been buffering global warming. The 'global dimming' that they have produced has reduced the amount of solar energy reaching the surface. The effect seems to be diminishing, however, so it has offered only a temporary respite from increased global temperatures.

9 | CLIMATE CHANGE

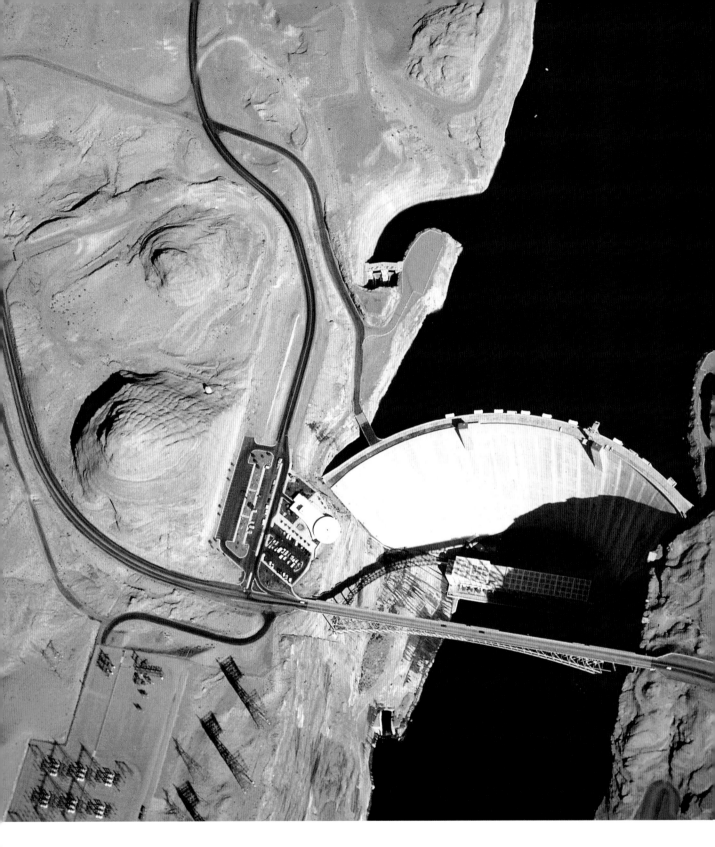

The Glen Canyon Dam

THIS DAM LIES ON THE COLORADO RIVER on the Arizona-Utah border in the USA, where it has created Lake Powell (top). It is 216 m/709 ft high, and one of the largest concrete dams in the world. The production of cement is one of the principal sources of carbon dioxide, and thus a major contributor to global warming. For every tonne of cement that is manufactured, one tonne of carbon dioxide is produced. In the United States alone, about 500 million tonnes of concrete are used every year, with cement forming between 10 and 15 per cent. That single source, for that one country alone, therefore releases between 50 and 75 million tonnes of carbon dioxide into the atmosphere every single year.

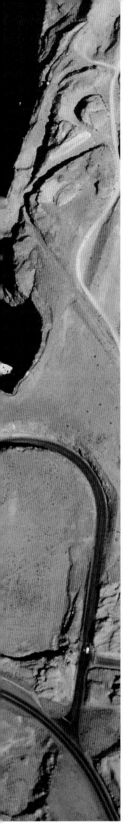

Previous pages | Icebergs calving from the Hubbard Glacier

MANY ICEBERGS ARE CREATED BY THE COLLAPSE of glaciers into the sea in the process known as calving. When the snout of a glacier reaches the sea and the ice starts to float, immense strains are placed on the terminal wall. The sections of ice that break away are the main source of large icebergs in the northern hemisphere, whereas sea ice is much thinner. The glaciers of Greenland, the northern Canadian islands, Spitzbergen, Novaya Zemlya, and (as here) Alaska are the main sources in the north. In the south, glaciers contribute icebergs from Patagonia and the Antarctic Peninsula, but giant tabular icebergs are also produced from the many ice shelves around Antarctica.

Concrete

251

CEMENT'S CONTRIBUTION TO GLOBAL WARMING does not end with its production. Quite apart from the indirect contribution made as it is transported long distances, it releases even more carbon dioxide when the chemical reactions occur that cause it to set. This coloured environmental scanning electron micrograph (ESEM) shows gypsum crystals (brown) that have formed in setting concrete (blue). Concrete consists of cement, water and aggregates such as sand and gravel. Adding water causes complex chemical changes to occur, hardening the cement around the aggregates, and it is at this point that the carbon dioxide is released.

Rice paddies and methane production

RICE PADDIES ARE A SURPRISING AGENT OF GLOBAL WARMING. Not an abstract painting, but an aerial view of extensive, terraced rice paddies in Richvale, California. The anaerobic (oxygen-deficient) conditions found in paddy fields (and other wetlands) mean that these are a major source of methane. (Other significant sources include ruminant animals and termites.) Although only traces are present in the atmosphere, its effect as a greenhouse gas is far greater than that of a similar quantity of carbon dioxide. Any increase in methane levels therefore has extremely serious consequences.

Termites and methane emissions

TERMITE MOUND IN MOSI O TUNYA, ZAMBIA. Termites are social insects that live in large colonies and build mounds that may be as much as 7 m/23 ft tall and have a diameter of 30 m/98 ft. Termites are one of the few living organisms that are able to break down the cellulose and lignin in dead wood through the action of specialized bacteria in their gut. In doing so they release nutrients into the soil (and thus make a major contribution to the local ecosystem). Unfortunately this process simultaneously releases large quantities of methane into the atmosphere.

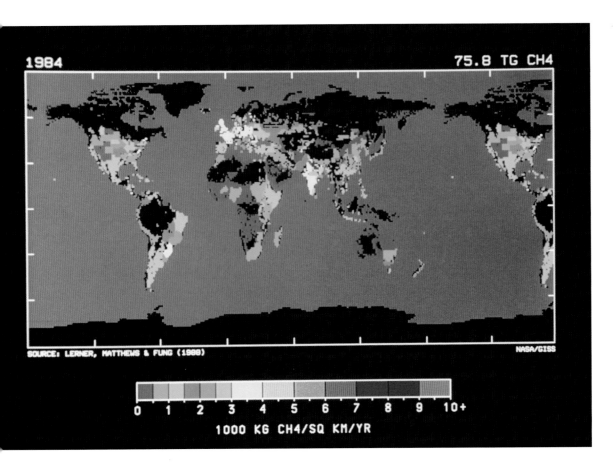

1984   75.8 TG CH4

SOURCE: LERNER, MATTHEWS & FUNG (1988)   NASA/GISS

0   1   2   3   4   5   6   7   8   9   10+

1000 KG CH4/SQ KM/YR

Methane emissions by animals

THE METHANE PRODUCTION BY DOMESTIC ANIMALS is shown in
this computer-generated map of the world. The map was created
by applying the estimated annual production of methane by specific animals (cattle, dairy cows, water buffalo, sheep, camels,
goats, pigs, horses and caribou) to their population. The scale
indicates tonnes of methane produced per square kilometre per
year. (Black areas show a lack of data.)

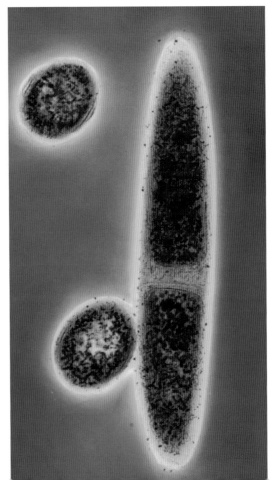

Methane and archaebacteria

ALTHOUGH CARBON DIOXIDE is the major component of global
warming, methane (CH_4), is not only a potent greenhouse gas, but
also contributes to ozone depletion. Although there are some geological sources of methane, most is produced by extremely primitive organisms – more primitive than bacteria – known as archaea,
here *Methanospirillum hungatii*. This is an anaerobic organism,
which lives in oxygen-free conditions, and metabolizes carbon
dioxide and hydrogen to produce methane. Unfortunately, the
effect of reducing the amount of carbon dioxide in the atmosphere
is far outweighed by the increased greenhouse effect from the
newly formed methane.

254

Hydroelectric power and global warming

CONTRARY TO ITS POPULAR IMAGE, HYDROELECTRIC POWER is not entirely 'green'. Apart from the environmental and human costs of construction, hydroelectric reservoirs are a source of methane. The deep water is often anaerobic (oxygen-poor) and forms a suitable environment for methanogenic organisms. In addition, drawing off water in dry seasons and subsequent refilling in rainy seasons means that the shorelines are also ideal habitats for methanogens. Reservoirs in tropical regions, such as this one at Catete in Angola, produce far greater quantities of methane than those in temperate zones where temperatures are much lower. The overall effect on global warming is often greater than a fossil-fuel power station producing the same amount of electricity.

255

Methane hydrate and iceworms

METHANE HYDRATE CONSISTS OF METHANE locked inside a matrix of water ice and occurs on the sea floor when low temperatures and high pressure are present. Vast quantities exist, and they are a potential source of energy greater than all known reserves of fossil fuels. There is the distinct possibility that slight warming of the oceans could lead to a sudden release of methane and a catastrophic increase in the greenhouse effect. This photograph of the sea floor, 700 m/2,297 ft deep in the Gulf of Mexico, shows that the hydrate deposits are home to large colonies of polychaete worms (iceworms), whose food supply may be closely linked to the hydrate itself.

Ice inclusions

THIS ICE SAMPLE COMES FROM AN ANTARCTIC CORE, drilled to a depth of 234 m/768 ft for the purpose of studying the concentration of methane in the past. Methane and other atmospheric gases are trapped within the bubbles found throughout the ice and seen here in polarized light. Radioactive dating techniques are used to determine the dates of the various layers, and in this case gave the sample's date as 1819. Analysis of the results shows that there was a steady rise in concentration from the early 1800s from 823 ppb (parts per billion) to 1481 ppb in 1978, since when the rate of increase appears to have begun to slow.

256

Sediment in Antarctic ice-core sample

THIS SAMPLE CAME FROM A RELATIVELY SHALLOW CORE from the permanently frozen surface of Lake Bonney, which lies in one of the Dry Valleys in Antarctica. The lower surface of the ice melts and refreezes over thousands of years, slowly transporting sediments from the surface down towards the lake. Far beneath the ice of East Antarctica lies Lake Vostok at a depth of 3,600 m/11,811 ft. It has been isolated from the atmosphere for hundreds of thousands of years, and could contain a unique microbial ecosystem. As with ice cores taken from the Greenland ice cap, deep cores taken from Antarctica reveal the past climate over many thousands of years.

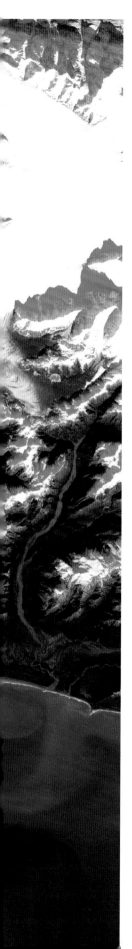

Retreat of the Bering Glacier

ALTHOUGH GLACIERS TEND TO ADVANCE AND RETREAT, depending on winter snowfall in their catchment areas, in recent years glaciers throughout the world have shown a consistent pattern of retreat. At first sight, these two images of the Bering Glacier, in Alaska (left: 1986; right: 2002), might appear to show little difference. Although the snout of the glacier remains in much the same place, it does show some slight changes. However, the main difference between these two images is that in the later one, the ice thickness has decreased considerably and more of the rock at the sides of the valleys is visible. This is particularly noticeable farther towards the source of the Bering Glacier (top right) and also alongside the smaller glaciers (top left).

Retreating Alpine glacier

GLACIERS CARVE OUT CHARACTERISTIC U-SHAPED VALLEYS as they erode the rocks along the sides of the valleys and from the surface beneath them. The rock debris is carried along by the flowing ice, and accumulates as lateral moraines at the sides of the glacier and a terminal moraine at its snout. The moraines remain when a glacier retreats. This glacier, in the Swiss Alps, has retreated considerably, leaving large lateral moraines and clearly showing that the sides of the valley have been scoured by what was once much thicker ice.

Glacial surges: Malaspina Glacier

THIS VAST GLACIER LIES WEST OF YAKUTAT BAY on the Canadian/Alaskan border. The highest peak in the background is Mt St Elias (5,489 m/18,009 ft), the second highest peak in Canada. No less than 25 glaciers combine here, forming the largest 'piedmont glacier' in the world, over 40 km/25 miles wide, and covering an area of 3,000 sq. km/1,158 sq. miles. The Malaspina Glacier is prone to unexpected surges, when meltwater collects beneath it, so that it is no longer frozen to its bed. The ice then surges forward and thins dramatically. Such glacial surges appear to be occurring more frequently and may be a sign of global warming.

Calving glacier

WHEN A GLACIER REACHES THE SEA, particularly as a result of a surge, it becomes unstable. The floating portion of the glacial tongue is subject to the rise and fall of the tides. The repeated flexure produces cracks even if no crevasses were present earlier, greatly weakening the ice. Large pieces of ice frequently fall into the sea in the process known as calving, as here at the Perito-Moreno Glacier, in Patagonia, in southern Argentina. This is one of the major sources of icebergs, although these tend to be relatively small, unlike the vast tabular icebergs that are sometimes produced by the break-up of the ice shelves around Antarctica.

Unstable ice sheets: Perutz Glacier

THIS GLACIER IS 16 KM/10 MILES LONG AND 3.2 KM/2 MILES WIDE and is one of the many found in West Antarctica. The stability of the ice sheet in West Antarctica is a source of concern. The majority of the glaciers on the Antarctic Peninsula are clearly in retreat, and others are unstable and subject to major surges. Over large areas, the base of the ice sheet is below sea level (unlike the situation in East Antarctica). Global warming could lead to the total collapse of West Antarctica's ice sheet, producing a sudden rise in sea level, flooding many highly populated regions and major cities around the world.

Giant tabular iceberg: B-15

THIS SATELLITE IMAGE SHOWS THE GIANT B-15 ICEBERG that broke away from the Ross Ice Shelf near Roosevelt Island in Antarctica in March 2000. At around 300 km/186 miles in length and 40 km/25 miles wide, it is one of the biggest icebergs ever known, although (at 12,000 sq. km/4,633 sq. miles) much smaller than the gigantic tabular berg seen off Scott Island in the Southern Ocean in November 1956, which was estimated to have an area of about 33,000 sq. km/12,741 sq. miles. Despite their size, such icebergs and the floating ice shelves around Antarctica do not produce any direct change in sea level when they melt.

The Maldives

ATOLLS CONSIST OF CORAL REEFS GROWING on submerged volcanoes. Within them are coral-sand islands that they protect from erosion. The Maldive atolls in the Indian Ocean consist of 26 atolls and 1,190 islands, all less than 2 m/ 6.6 ft above current sea level. Even if they recover from the devastation caused by the tsunami of 26 December 2004, the Maldives, with the island chains of Kiribati and Tuvalu in the western Pacific, are likely to be the first casualties of the rising sea levels that will inevitably accompany global warming.

Chilean fjords

DURING THE LAST ICE AGE, glaciers gouged out their typical U-shaped valleys in many areas, including Norway, the South Island of New Zealand, and (as here) in southern Chile. When sea-levels rose, the valleys flooded to give typical deep fjords with steep sides, and sometimes an abrupt underwater step or bar at the seaward mouth. Their profile is unlike that of rias (as in western Ireland) which were cut by water rather than ice, and which are V-shaped in cross-section with more gently sloping valley sides.

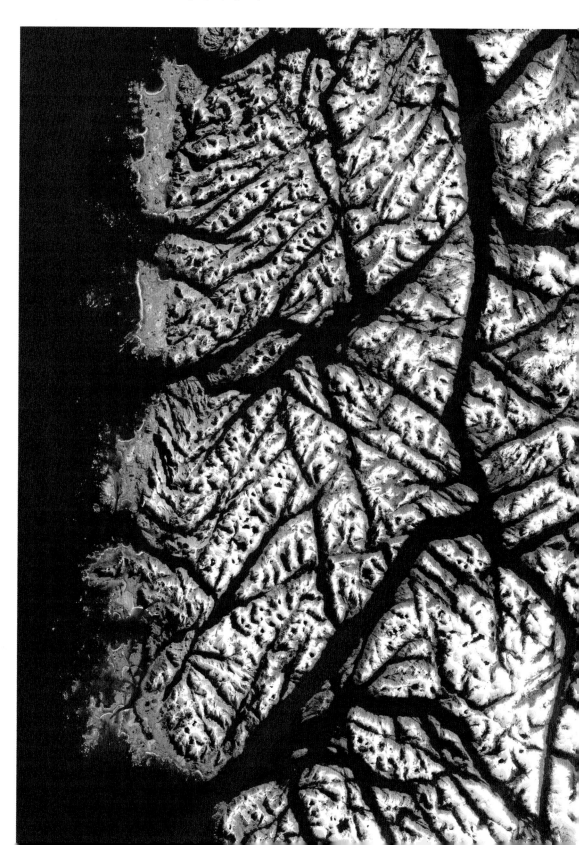

Bleached coral

CORAL CONSISTS OF MILLIONS OF TINY, INDIVIDUAL POLYPS, which are only able to survive in the presence of certain symbiotic algae (*zooxanthellae*). Death of the algae results in the death of the coral. Only the coral's calcium-carbonate skeleton remains, producing an obvious loss of colour (bleaching). The exact trigger for the death and bleaching is uncertain, although pollution and, more especially, increasing sea temperatures are important factors. Dead coral is obviously unable to regenerate after damage by storms or tsunamis, so global warming brings the risk that there will be widespread loss of coral reefs, which currently protect neighbouring shorelines from damage.

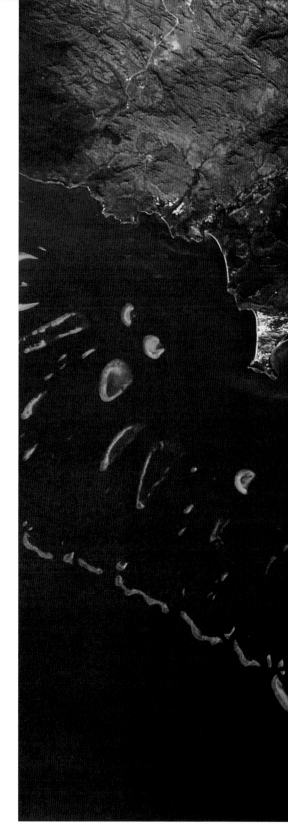

Fan coral on a healthy reef

IT HAS RECENTLY BEEN DISCOVERED THAT, strangely, corals may not be defenceless against global warming and rising sea temperatures. As part of their normal metabolism, corals produce a gas known as dimethyl sulphide (DMS). This escapes into the atmosphere, where it provides extremely effective condensation nuclei for cloud droplets. There is the distinct possibility that with rising temperatures there will be a rise in the production of DMS, and a corresponding increase in cloudiness over tropical oceans, with resultant cooling of the surface waters. This may provide some degree of protection for corals against the worst effects of global warming.

Great Barrier Reef and coast of Australia

THIS IMAGE SHOWS PRINCESS CHARLOTTE BAY (top right) in northern Queensland, Australia. North is to lower right. The reefs form shallow seas and small islands (white/green) surrounded by deeper water (dark blue). The land (brown) is mostly arid or semi-arid. The outer reefs (ribbon reefs) form a chain, 40–80 km/25–50 miles out from the shore, along the edge of the continental shelf, a shallower area between land and the ocean depths. If global warming does destroy such fringing reefs, the warm waters close to the land will be replaced by cooler oceanic water. This will almost certainly result in a change in climate. Depending on the precise local conditions and prevailing winds, rainfall is likely to be reduced, causing the land to become even more arid.

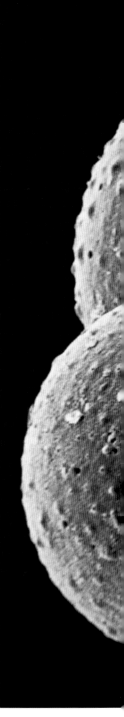

Global atmospheric archive

ONE OF THE PROBLEMS FACING climatologists is that records of the concentrations of gases and pollutants in the atmosphere do not extend very far into the past. To try to remedy this situation – at least for the future – atmospheric samples, collected at the Cape Grim Baseline Air Pollution Station at the north-western tip of Tasmania, are here seen stored in pressurized gas cylinders at the Global Atmospheric Sampling Laboratory (GASLAB) in Melbourne, Australia. The archive may be used to check concentrations of gases that were not monitored in the past, or to refine existing measurements as analytical techniques improve.

Deep bore Vostok ice-core samples

CORES HAVE BEEN RECOVERED from a depth of 3,350 m/2,082 miles below Antarctica that seem to date to 400,000 years ago. Important information may be obtained about the prevailing temperatures throughout that period from determination of the ratios of oxygen isotopes that are present. Water molecules (H_2O) containing the lighter isotope, ^{16}O are preferentially evaporated from the oceans, which are correspondingly enriched in ^{18}O. The Antarctic and Greenland ice caps, however, are enriched in ^{16}O and deficient in ^{18}O. The fluctuations in the ratio of the two isotopes are temperature-dependent, enabling past temperatures to be determined. Worryingly, the cores show that abrupt, major changes in climate have occurred in just decades, rather than over centuries as had been expected.

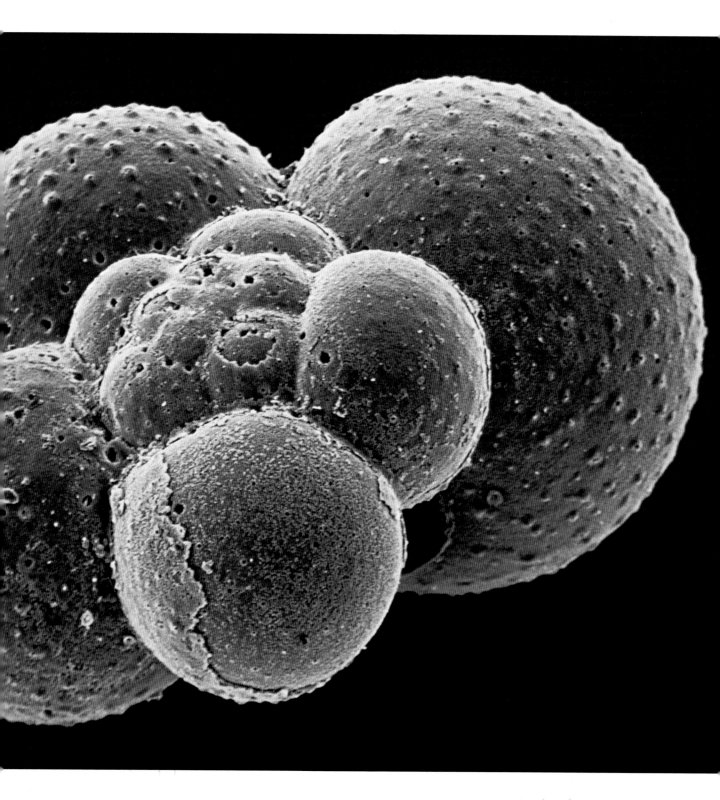

Marine foraminafera

FORAMINAFERA ARE SINGLE-CELLED ORGANISMS that are widespread throughout the oceans and are a valuable indicator of past climates. In some species that occupy multi-chambered shells or 'tests', the direction in which they coil is closely dependent upon the ambient temperature, altering at higher or lower values. Although this provides some information about prevailing temperatures, even more important information is derived from an analysis of the ratio of the two oxygen isotopes, ^{16}O and ^{18}O, in the calcium carbonate ($CaCO_3$) of tests recovered from sea-floor cores. During ice ages the polar ice caps contain particularly large quantities of ^{16}O, and foraminafera tests, formed from sea-water, are greatly enriched in ^{18}O.

Air pollution, northern India, Bangladesh and the Bay of Bengal

THE HAZE CONTAINS A MIXTURE of liquid and solid particles generated by industry, vehicle emissions and domestic fires. Worldwide, such pollution is so widespread that, together with slash and burn agriculture and other sources of smoke and haze, it is largely responsible for 'global dimming' – a reduction in the amount of sunlight that reached the Earth's surface in recent decades. This partially offset any effects of global warming, but the dimming now seems to be decreasing. Such pollution must be further reduced, but this is likely to lead to an increase (possibly a dramatic increase) in global warming. Both issues must be tackled urgently and simultaneously.

Aircraft contrails over the Central Rhône Valley, France

272 THIS PHOTOGRAPH ILLUSTRATES ANOTHER FACTOR to be taken into account in modelling climate change. There is some cirrus cloud in the west (left), but most of the cloud consists of condensation trails (contrails), which form when the atmosphere is humid. (In dry air contrails evaporate quickly.) Persistent contrails reflect sunlight back into space, cooling the surface below them. Climate models not only need to include realistic models for contrails (and all clouds) but also reliable estimates of future levels of air travel. In this image, Geneva and Lake Geneva in Switzerland are at top centre. The border between Switzerland and Italy lies between the two snow-free valleys at top right (the headwaters of the Rhône in the north and the Val d'Aosta in the south).

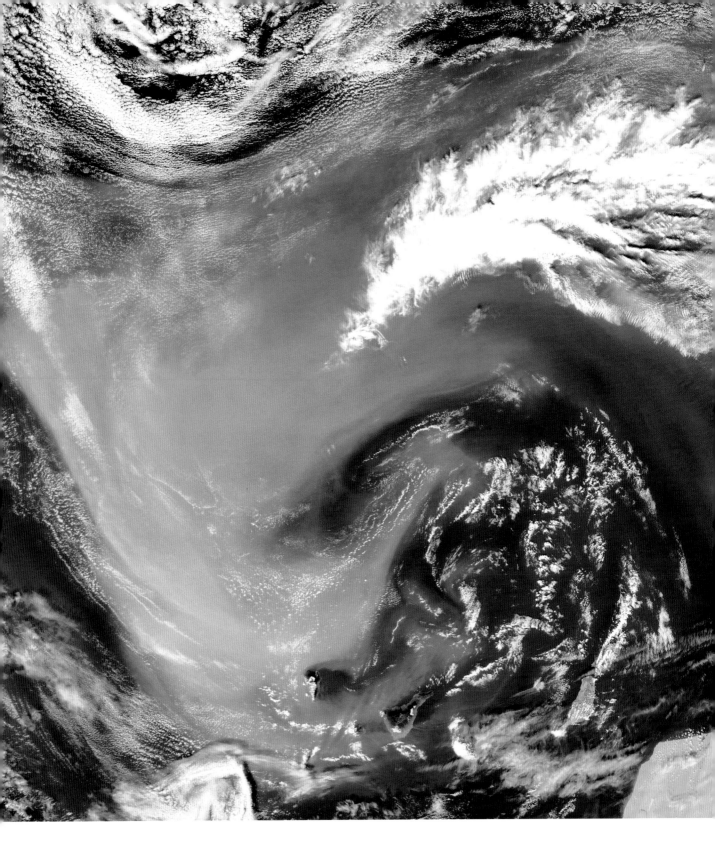

Saharan dust over Atlantic

DUST FROM THE SAHARA HAS BEEN CARRIED ROUND the northern side of a depression by the surface winds, and is affecting the Canary Islands (bottom centre). Fine dust from the Sahara often spreads right across the Atlantic and makes a significant contribution to haze and particulate levels over North and South America. The dust actually provides valuable nutrients to the Amazonian rainforest. If global warming produces greater rainfall over the Sahara, with a return of the extensive vegetation that once covered the area, dust storms are likely to weaken, or cease completely, accelerating the decline of the rainforest on the other side of the ocean.

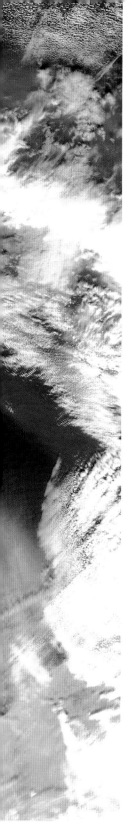

Beneath desert sands

THE BLACK-AND-WHITE STRIP IS A RADAR IMAGE of part of the Sahara Desert in southeast Egypt. It is superimposed on an image obtained at a visible wavelength. The radar pulses penetrate the surface deposits of dry, windblown sand and reveal features in the bedrock below, including drainage channels, cut when the area had a much wetter climate. Such a technique is possible only over arid regions, because moisture in the soil reduces the depth to which the radio waves penetrate.

Ash plume, Mt Pinatubo

A FALSE-COLOUR IMAGE, TAKEN ON 11 JUNE 1991, of the plume of volcanic ash from Mt Pinatubo on the island of Luzon in the Philippines. The yellow areas indicate normal clouds. Quite apart from the effects of ashfall, lava flows and pyroclastic clouds, eruptions frequently exhibit some effect on the weather in the surrounding region. Areas under the plume suffer lower temperatures, and aerosols act as condensation nuclei, frequently leading to heavy rain, which can produce highly destructive mudflows (known as lahars) as the unconsolidated volcanic deposits are swept away by the rain. The area around Pinatubo suffered greatly from lahars when a tropical cyclone and its accompanying torrential rainfall moved in from the Pacific.

276

WEATHER

Plinian column, Mt St Helens

VOLCANOES THAT HAVE PLINIAN-TYPE ERUPTIONS eject vast quantities of material, much of it as a vertically directed column of gas, fragments of lava and pulverized rock. In this eruption of 22 July 1980 (after the initial catastrophic blast of 18 May) the column reached a height of about 12 km/ 7.5 miles. Other eruption columns, such as that of Mt Pinatubo in the Philippines in 1991, reach 20 km/12.4 miles or even more, sending aerosols (finely divided liquid droplets or solid particles) well into the stratosphere.

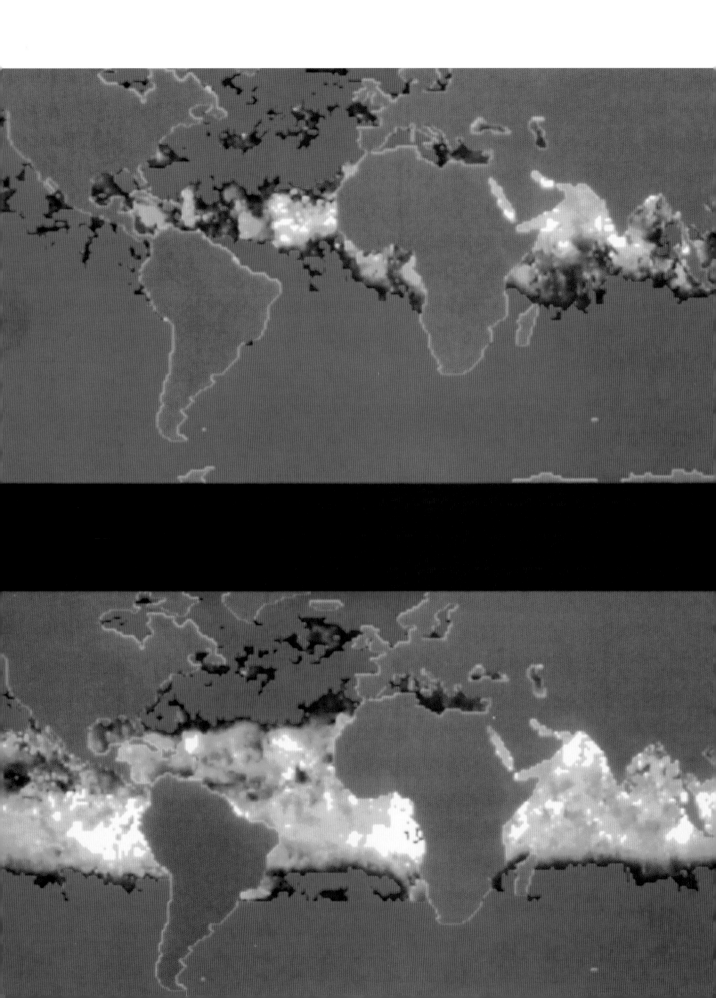

Global effects of eruption: Mt Pinatubo

THESE TWO FALSE-COLOUR MAPS show the distribution of volcanic aerosols in the atmosphere, immediately after the eruption of Mt Pinatubo (top) and two months later (bottom). Aerosol concentration is shown as shades of yellow, from brown (lowest) to white (highest). Mt Pinatubo, in the Philippines, erupted in June 1991, culminating in a vast explosion on 15–16 June. The top image (19–27 June) shows a near-normal aerosol distribution, with only a slight increase over the Indian Ocean. The bottom image (8–14 August) shows a vast aerosol plume around the Equator.

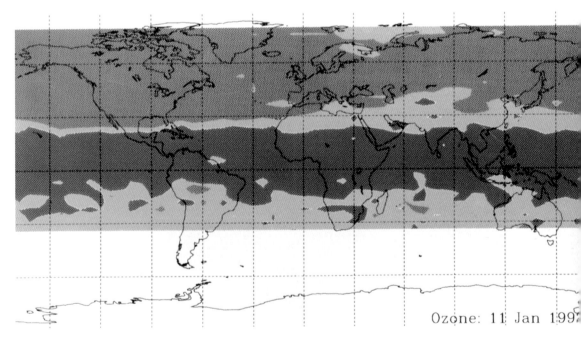

Ozone: 11 Jan 1992

Ozone depletion following volcanic eruption

THIS MAP SHOWS THE NORTHERN-HEMISPHERE and tropical distribution of stratospheric ozone for 11 January 1992, following the eruption of Mt Pinatubo in June 1991. The concentration is colour-coded from blue (lowest) to red (highest). The large loss of ozone over the tropics coincides roughly with the plume from Mt Pinatubo, suggesting that sulphur compounds in the volcanic aerosols act to catalyze ozone depletion. The depletion in the tropics was about 10 per cent at low levels and as high as 50 per cent at an altitude of 20 km/12.4 miles.

Caldera of Mt Tambora, Sumbawa Island, Indonesia

THE CALDERA (6 km/4 miles wide and 650 m/2,133 ft deep) was formed in 1815 by the greatest eruption of historic times, far greater than the better-known eruption of Krakatau in 1883. Aerosols from Tambora's eruption were ejected high into the stratosphere and were produced in such vast quantities that they not only affected the tropics but also blanketed the whole Earth. They caused a dramatic reduction in the amount of solar energy reaching the surface and had an immediate effect on the weather. The winter of 1815–1816 was exceptionally cold and persistent, and both spring and summer in 1816 were abnormally cold, so that 1816 became known as `the year without a summer'. Crop failures and even famine conditions were widespread.

Greatest known volcanic eruption

THIS IMAGE SHOWS THE ENORMOUS CALDERA OF MT TOBA IN SUMATRA, left by the greatest eruption known to volcanologists. Lake Toba occupies an area of about 1,300 sq. km/502 sq. miles, including Samosir Island which is about 30 km/19 miles long by 10 km/6 miles wide. The gigantic eruption occurred in prehistoric times, estimated to have been around 74,000 BP (Before Present). The amount of ejecta is incredible: it appears to have been about 3,000 cubic km/720 cubic miles. For comparison, the volume of ejecta from Mt Tambora was about 50 cubic km/12 cubic miles, and from Mt St Helens a mere 3 cubic km/0.7 cubic miles. Such a gigantic eruption must have had a devastating effect on the Earth's weather.

GLOSSARY
&INDEX

GLOSSARY

ANTICYCLONE

A high-pressure region where air subsides from higher altitudes, and then flows out over the surrounding area. The circulation around an anticyclone is clockwise in the northern hemisphere.

ANTISOLAR POINT

The point on the sky directly opposite the position of the Sun, relative to the observer's head.

CELL

A volume of air within a cloud, in which convection occurs largely independently of activity in adjoining cells. Cells go through stages of growth, maturity, and decay. Large clouds consist of many cells at different stages of development, and multicell thunderstorms may contain several cells with lightning activity. Within major cumulonimbus clusters cells sometimes evolve into a single, giant cell, creating a supercell storm, which produces violent weather and may spawn tornadoes.

CONDENSATION LEVEL

The level in the atmosphere at which water vapour condenses into cloud droplets. The exact level (or levels) depends upon the humidity (water content) of the air, and the temperature of its surroundings.

CONTINENTAL CLIMATE

A climate typical of continental interiors, characterized by extremely cold winters and hot summers. Such areas also tend to have low average annual rainfall.

CYCLONE

A system in which air circulates around a low-pressure centre, with two distinct meanings: 1) a 'tropical cyclone', a self-sustaining tropical storm, also known as a hurricane or typhoon; 2) an 'extratropical cyclone' or depression, a low-pressure area, one of the main weather systems affecting temperate regions.

DEPRESSION

The term generally used for a low-pressure area. Air flows into a depresssion and rises in its centre. Known technically as an 'extratropical cyclone'. The wind circulation around a depression is cyclonic (anticlockwise in the northern hemisphere, clockwise in the southern).

FIRN

Snow that has persisted for some time, during which it has altered in texture, often by melting and refreezing, to become granular in nature, and is in the process of becoming glacier ice.

HURRICANE

One of several names for a potentially destructive tropical cyclone, used in the North Atlantic and eastern Pacific.

INSTABILITY

The condition under which a parcel of air, if displaced upwards or downwards, tends to continue (or even accelerate) its movement. It is the opposite of stability.

INVERSION

An atmospheric layer in which temperature increases with height.

ISOBAR

A line that joins points on a weather chart that have the same barometric pressure.

JET STREAM

A narrow band of high-speed winds that lies close to a major boundary between air with markedly different temperatures. There are two important jet streams in each hemisphere (the polar-front and sub-tropical jet streams). There are other jet streams in the tropics and at high altitudes.

MARITIME CLIMATE

A climate that is strongly influenced by the region's proximity to the ocean. Generally characterized by significant amounts of precipitation throughout the year, but with generally mild winters and summers that rarely reach extremely high temperatures.

MESOSPHERE

The atmospheric layer above the stratosphere, within which temperature decreases with height, reaching the atmospheric minimum at the mesopause, at an altitude of about 86 or 100 km (depending on season and latitude).

OROGRAPHIC CLOUD

Any cloud that has been produced by the uplift caused by the air being forced to rise over hills or mountains.

OZONE

Triatomic oxygen, O_3, a highly reactive, colourless gas. It is naturally present in the stratosphere, where it absorbs otherwise damaging ultraviolet light, and causes a rise in temperature in that atmospheric layer. Its increasing concentration in the upper troposphere, where it is created by aircraft exhaust gases, contributes to global warming. At low level, it is a pollutant and arises primarily from vehicle exhausts.

OZONE HOLE

A region in which stratospheric ozone is greatly depleted, allowing harmful ultraviolet radiation to reach the ground. The ozone is destroyed in chemical reactions, particularly with the compounds known as chlorofluorocarbons (CFCs), the use of which has now been banned worldwide. Ozone holes develop with the return of sunlight (which promotes the chemical reactions) to polar regions in the spring.

PARHELION

The technical term for a mock sun.

PRECIPITATION

The technical term for water in any liquid or solid form that is deposited from the atmosphere, and which falls to the ground. It excludes cloud droplets, mist, fog, dew, frost and rime, as well as virga (trails of rain, snow, or ice crystals that do not reach the ground).

RADIOMETER

The principal instrument carried by many meteorological satellites. Radiometers can receive radiation from the Earth over a wide spectral range, often from the ultraviolet to the infrared. Normally, however, several filters are used to divide coverage into narrow bands, which provide information about specific properties of the surface or atmosphere. One such channel is frequently chosen to give coverage similar to the visual response of the human eye.

STRATOSPHERE

The second major atmospheric layer from the ground, in which temperature initially remains constant, but then increases with height. It lies between the troposphere and the mesosphere, with a lower boundary of approximately 8–20 km (depending on latitude) and an upper one of about 50 km.

STABILITY

The condition under which a parcel of air, if displaced upwards or downwards, seeks to return to its original position rather than continuing its movement.

SUPERCELL THUNDERSTORM

A giant cumulonimbus complex, within which the updraughts have become organized into a single, rotating updraught, with various compensating downdraughts and outflows. Supercells have a long lifetime of several hours and often produce violent lightning activity, destructive hail, torrential rain and major tornadoes.

SUPERCOOLING

The conditions under which water may exist in a liquid state, despite being at a temperature below 0°C. This occurs frequently in the atmosphere, often in the absence of suitable freezing nuclei. Even without freezing nuclei, supercooled water will freeze if the temperature drops to -40°C/-40°F.

THERMAL

A rising bubble of air that has broken away from the heated surface of the ground. Depending on the exact conditions, a thermal may rise until it reaches the condensation level, at which its water vapour will condense into droplets, giving rise to a cloud.

THERMOCLINE

A layer in which there is an abrupt change in water temperature, particularly the boundary between warm surface water and cool deep oceanic water.

TROPICAL STORM

The stage immediately preceeding (or following) a full-scale tropical cyclone. There is a well-developed circulation around a low-pressure centre, and wind speeds may reach 8–11 on the Beaufort scale (62 kph/38 mph to 117 kph/73 mph).

TROPOPAUSE

The inversion that separates the troposphere from the overlying stratosphere. Its altitude varies from approximately 8 km at the poles to 18–20 km over the equator.

TROPOSPHERE

The lowest region of the atmosphere in which most of the weather and clouds occur. Within it, there is an overall decline in temperature with height.

WAVE CLOUDS

Clouds that arise as the result of wave-motion of the air, usually caused by its passage over hills or mountains on the surface. The clouds form at the crests of the otherwise invisible waves, where the air is lifted above its condensation level, but generally decay where the air descends into the succeeding trough. Trains of wave clouds may form downwind of the initial obstacle.

WIND SHEAR

A change in wind direction or strength with a change of position. The most common occurrence is when wind strength increases with increasing height above the ground.

ZENITH

The point on the sky directly above the observer's head.

INDEX

CREDITS

1 BRITISH ANTARCTIC SURVEY /SCIENCE PHOTO LIBRARY; 2–3 NASA /SPL; 8–9 ART WOLFE /SPL; 10 DAVID HAY JONES /SPL; 11 DAVID HAY JONES /SPL; 12–13 PEKKA PARVIAINEN /SPL; 14 HANS NAMUTH /SPL; 15 PEKKA PARVIAINEN /SPL; 16 PEKKA PARVIAINEN /SPL; 17 PEKKA PARVIAINEN /SPL; 18–19 DAVID PARKER /SPL; 20 JIM REED /SPL; 21t PEKKA PARVIAINEN /SPL; 21b PEKKA PARVIAINEN /SPL; 22 ROBIN SCAGELL /SPL; 23 PEKKA PARVIAINEN /SPL; 24 VAUGHAN FLEMING /SPL; 25 NASA /SPL; 26 NASA /SPL; 27 NASA /SPL; 28–29 DOUG ALLAN /SPL; 30 PEKKA PARVIAINEN /SPL; 31 PEKKA PARVIAINEN /SPL; 32 GORDON GARRADD /SPL; 33 ROGER APPLE-TON /SPL; 34t PEKKA PARVIAINEN /SPL; 34b PEKKA PARVIAINEN /SPL; 35 NASA /SPL; 36–37 GEOFF TOMPKIN-SON /SPL; 38 NASA /SPL; 39t SIMON FRASER /SPL; 39b STEVE PERCIVAL /SPL; 40–41 JON DAVIES /JIM REED PHOTOGRAPHY /SPL; 42 GORDON GARRADD /SPL; 43 JIM REED /SPL; 44 MAGRATH /FOLSOM /SPL; 45 BRITISH ANTARCTIC SURVEY /SPL; 46 NASA /SPL; 47 NASA /SPL; 48–49 MIKE BOYATT / AGSTOCK /SPL ; 50 GEORGE POST /SPL; 51 DR JEREMY BURGESS /SPL; 52–53 ASTRID & HANNS-FRIEDER MICHLER /SPL; 54 SIMON FRASER /SPL; 55r ROD PLANCK /SPL; 55l © Richard Fleet 2004; 56t DR JUERG ALEAN /SPL; 56b SIMON FRASER /SPL; 57 MAGRATH PHOTOGRAPHY /SPL; 58 DR JEREMY BURGESS /SPL; 59 PEKKA PARVIAINEN /SPL; 60–61 SIMON FRASER /SPL; 62 CNES, DISTRIBUTION SPOT IMAGE /SPL; 63 HASLER & PIERCE, NASA GSFC /SPL; 64t CNES, 1986 DISTRIBUTION SPOT IMAGE /SPL; 64b NASA /SPL; 65 M-SAT LTD /SPL; 66 CNES, DISTRIBUTION SPOT IMAGE /SPL; 67 2002 ORBITAL IMAGING CORPORATION /SPL; 68 CNES, 1998 DISTRIBUTION SPOT IMAGE /SPL; 69 CNES, 1998 DISTRIBUTION SPOT IMAGE /SPL; 70–71 GEORGE RANALLI /SPL; 72 G. ANTONIO MILANI /SPL; 73 ADAM JONES /SPL; 74–75 ORBIMAGE /SPL; 76 ROBERT BROOK /SPL; 77 NRSC LTD /SPL; 78 PETER CHADWICK /SPL; 79 BRITISH ANTARCTIC SURVEY /SPL; 80 2002 ORBITAL IMAGING CORPORATION /SPL; 81 NASA /SPL; 82 NASA /SPL; 83 NASA /SPL; 84 ADAM HART-DAVIS /SPL; 85 NASA /SPL; 86 2002 ORBITAL IMAGING CORPORATION /SPL; 87 DAVID R. FRAZIER /SPL; 88–89 GEORGE HOLTON /SPL; 90tl TED KINSMAN /SPL; 90tr KENNETH LIBBRECHT /SPL; 90b KENNETH LIBBRECHT /SPL; 91cl TED KINSMAN /SPL; 91tr TED KINSMAN /SPL; 91tl KENNETH LIB-BRECHT /SPL; 91bl KENNETH LIBBRECHT /SPL; 91cr KENNETH LIBBRECHT /SPL; 92 JIM REED /SPL; 93b ASTRID & HANNS-FRIEDER MICHLER /SPL; 93t NCAR /SPL; 94 ADRIENNE HART-DAVIS /SPL; 95 SIMON FRASER /SPL; 96 A.C. TWOMEY /SPL; 97tr JIM REED /SPL; 97bl LARRY WEST /SPL; 98 BERNHARD EDMAIER /SPL; 99 BERNHARD EDMAIER /SPL; 100t BRITISH ANTARCTIC SURVEY /SPL; 100b SIMON FRASER /SPL; 101 BERNHARD EDMAIER /SPL; 102 BERNHARD EDMAIER /SPL; 103 B&C ALEXANDER /SPL; 104 BRITISH ANTARCTIC SURVEY /SPL; 105 BERNHARD EDMAIER /SPL; 106 SIMON FRASER /SPL; 107 NASA /SPL; 108 CNES, 1989 DISTRIBUTION SPOT IMAGE /SPL; 109 SIMON FRASER /SPL; 110 BERNHARD EDMAIER /SPL; 111 SIMON FRASER /SPL; 112–113 JIM REED /SPL; 114 NASA /SPL; 115 FRED K. SMITH /SPL; 116 KEITH KENT /SPL; 117 KEITH KENT /SPL; 118 DAVID PARKER /SPL; 119 PETER MENZEL /SPL; 120 CLEM HAAGNER /SPL; 121 J.G. GOLDEN /SPL; 122 G. BRAD LEWIS /SPL; 123 JIM REED /SPL; 124 JIM REED /SPL; 125 REED TIMMER & JIM BISHOP /JIM REED PHOTOGRAPHY /SPL; 126 AARON JOHNSON & BROOKE TABOR / JIM REED PHOTOGRAPHY /SPL; 127 AARON JOHNSON & BROOKE TABOR / JIM REED PHOTOGRAPHY /SPL; 128 REED TIMMER AND JIM BISHOP / JIM REED PHOTOGRAPHY /SPL; 129 NASA / GODDARD SPACE FLIGHT CENTER/SPL; 130 SPL; 131 NASA /SPL; 132 NASA /SPL; 133 NASA /SPL; 134 CHRIS SATTLBERGER /SPL; 135 CHRIS SATTLBERGER /SPL; 136 NASA / GODDARD SPACE FLIGHT CENTER /SPL; 137 NASA /SPL; 138 NASA /SPL; 139t CHRIS SATTLBERGER /SPL; 139b JIM REED /SPL; 140–141 SIMON FRASER /SPL; 142 NRSC LTD /SPL; 143 NASA /SPL; 144 NASA /SPL; 145 NASA /SPL; 146–147 ROY L. BISHOP / AMERICAN INSTITUTE OF PHYSICS /SPL; 148 MARK A. SCHNEIDER /SPL; 149 PEKKA PARVIAINEN /SPL; 150 GOR-DON GARRADD /SPL; 151 FRANK ZULLO /SPL; 152–153 PEKKA PARVIAINEN /SPL; 155 PEKKA PARVIAINEN /SPL; 156 MICHAEL GIANNECHINI /SPL; 157 SIMON FRASER /SPL; 158–159 JOHN FOSTER /SPL; 160 DR MORLEY READ /SPL; 161 GORDON GARRADD /SPL; 162–163t GEORGE POST /SPL; 162–163b FRANK ZULLO /SPL; 164–165 PEKKA PARVIAINEN /SPL; 166 CHRIS DAWE /SPL; 167 DAVID NUNUK /SPL; 168–169 FRANK ZULLO /SPL; 170–171t DAMIEN LOVEGROVE /SPL; 170–171b PEKKA PARVIAINEN /SPL; 172 PEKKA PARVIAINEN /SPL; 173 PEKKA PARVI-AINEN /SPL; 174 PEKKA PARVIAINEN /SPL; 175 PEKKA PARVIAINEN /SPL; 176 PEKKA PARVIAINEN /SPL; 177 PEKKA PARVIAINEN /SPL; 178 CHRIS MADELEY /SPL; 179 JACK FINCH /SPL; 180 LIONEL F. STEVENSON /SPL; 181 NASA /SPL; 182 NASA /SPL; 183 PEKKA PARVIAINEN /SPL; 184 John Hardwick; 185 SIMON FRASER /SPL; 186–187 NASA /SPL; 188t NOAA /SPL; 188b © 2005 EUMETSAT; 189 R.B.HUSAR / NASA /SPL; 190 EUROPEAN SPACE AGENCY /SPL; 191t EUROPEAN SPACE AGENCY /SPL; 191b EUROPEAN SPACE AGENCY /SPL; 192 ESA / PLI /SPL; 193 PHOTO LIBRARY INTERNATIONAL / ESA /SPL; 194 UNIVERSITY OF DUNDEE /SPL; 195 NASA/Jeff Schmaltz, MODIS Land Rapid Response Team; 196 NASA /SPL; 197 NASA /SPL; 198 M-SAT LTD /SPL; 199 NASA /SPL; 200–201 NASA /SPL; 202 NASA /SPL; 203 NASA /SPL; 204 JIM REED /SPL; 205 NOAA /SPL; 206 DAVID PARKER /SPL; 207 PEKKA PARVIAINEN /SPL; 208 NOAA /SPL; 209 NOAA /SPL; 210t NASA /SPL; 210b NASA /SPL; 211t NASA /SPL; 211b NASA /SPL; 212–213t NASA /SPL; 212–213b NASA /SPL; 214–215 NASA GSFC /SPL; 216 LAWRENCE MIGDALE /SPL; 217 NASA /SPL; 218 NRSC LTD /SPL; 219 2002 ORBITAL IMAGING CORPORATION /SPL; 220–221 J.G. PAREN /SPL; 222t US GEOLOGICAL SURVEY /SPL; 222b NASA /SPL; 223 M-SAT LTD /SPL; 224–225 NASA /SPL; 226 BP / NRSC /SPL; 227 SIMON FRASER /SPL; 228 SIMON FRASER /SPL; 229 BERNHARD EDMAIER /SPL; 230 Courtesy of Natural Resources Canada, photo no. A89S0052; 231 ALAN SIRULNIKOFF /SPL; 232 ALAN SIRUL-NIKOFF /SPL; 233 NASA /SPL; 234 ANDREW BROWN /SPL; 235 CHRIS SATTLBERGER /SPL; 236 GEOFFREY S CHAPMAN /SPL; 237 JOHN READER /SPL; 238 DR MORLEY READ /SPL; 239b JOHN READER /SPL; 239t DR MOR-LEY READ /SPL; 240 BRIAN BRAKE /SPL; 241 Chen Su © Panorama Stock Photos Co Ltd /Alamy; 242 CNES, 1986 DISTRIBUTION SPOT IMAGE /SPL; 243 EARTH SATELLITE CORPORATION /SPL; 244 BERNHARD EDMAIER /SPL; 245 EARTH SATELLITE CORPORATION /SPL; 246–247 DOUG ALLAN /SPL; 248–249 BERNHARD EDMAIER /SPL; 250 GARY LADD /SPL; 251 PASCAL GOETGHELUCK /SPL; 252 PETER MENZEL /SPL; 253 PETER CHADWICK /SPL; 254t NASA, GODDARD INSTITUTE FOR SPACE STUDIES/SPL; 254b DR KARI LOUNATMAA /SPL; 255 CNES, 1989 DISTRIBUTION SPOT IMAGE /SPL; 256t Photo courtesy of Ian R. MacDonald, Texas A&M Univ. Corpus Christi; 256b CSIRO /SPL; 257 ED ADAMS / MONTANA STATE UNIVERSITY /SPL; 258 NASA /SPL; 259 NASA /SPL; 260–261 JERE-MY WALKER /SPL; 262 BERNHARD EDMAIER /SPL; 263 BERNHARD EDMAIER /SPL; 264 DAVID VAUGHAN /SPL; 265 NASA /SPL; 266 ALEXIS ROSENFELD /SPL; 267 M-SAT LTD /SPL; 268t ALEXIS ROSENFELD /SPL; 268b MATTHEW OLDFIELD, SCUBAZOO /SPL; 269 NASA /SPL; 270b MUNOZ-YAGUE /EURELIOS /SPL; 270t CSIRO / SIMON FRASER /SPL; 271 MANFRED KAGE /SPL; 272 NASA /SPL; 273 NASA /SPL; 274 NASA /SPL; 275 NASA /SPL; 276 ROBERT M CAREY, NOAA /SPL; 277 PROF. STEWART LOWTHER /SPL; 278 ROBERT M CAREY, NOAA /SPL; 279 NASA /SPL; 280 NASA /SPL; 281 USGS, PHOTOGRAPH BY THOMAS J. CASADEVALL; 282–283 JIM REED /SPL

A CASSELL ILLUSTRATED BOOK

An Hachette Livre UK Company

First published in Great Britain in 2006 by Cassell Illustrated,

This edition published in Great Britain in 2007 by Cassell Illustrated, a division of Octopus Publishing Group Limited, 2–4 Heron Quays, London, E14 4JP

Text Copyright Storm Dunlop

Design and Layout © 2006 Octopus Publishing Group Limited

The moral right of Storm Dunlop to be identified as the author of this Work has been asserted in accordance with the Copyright, Designs and Patents Act of 1988

ISBN-10: 1 84403 601 4
ISBN-13: 9781844036011

Commissioning Editor: Karen Dolan
Project Editor: Joanne Wilson
Editor: Katie Hewett
Design: Austin Taylor
Index: Sue Bosanko

Printed in China

Special thanks to Kevin Davis at the Science Photo Library.

Caption for page 282–283

METEOROLOGISTS FROM THE SEVERE THUNDERSTORM Electrification and Precipitation Study (STEPS) launch a weather balloon into a storm in Kansas U.S.A. This is a tornadic supercell thunderstorm, a massive storm in the process of developing a tornado. The weather balloon carries instruments to obtain data on temperature, pressure, wind speed and electrical fields within storm.